Disclosing Elite Ecologies

Disclosing Elite Ecologies: Methodologies For "Doing" Urba[illegible]ers a set of methodologies to chart urban elites. Whereas most researc[illegible] has focused on the global super-rich, this book pays specific attention to the multidimensional urban geographies of elite reproduction and transformation, as elites depend on urban contexts for capital accumulation, consumption and leisure, and housing.

Adopting an interdisciplinary approach to the topic, contributing authors discuss various theoretical and methodological antecedents in urban studies and related areas of research that have investigated economic elites. Building on, but also moving beyond these bodies of literature, the book rejects a-priori definitions of the size and shape of this social group and instead pursues relational, place specific conceptualizations of elite composition and behavior. In particular, the contributions to the volume show that urban elite research benefits from paying more attention to: (i) boundary work between elites and non-elites; (ii) intra-elite competition and distinction; (iii) national state spaces in determining elite composition; and (iv) the urban sense of belonging of economic elites.

This extensive volume provides readers with various empirical inroads into the study of urban elites drawing on research set in Brussels, Fez, London, Lyon, Madrid, Manchester, Milan, New York City, Paris, and Porto Alegre. Taking inspiration from urban and economic geography, elite theory and urban sociology, cultural sociology, political economy, anthropology, criminology, architecture, and migration studies, this book aims to open up the opportunity for methodological cross-fertilization.

The chapters in this book were originally published as a special issue of *Urban Geography*.

Bas van Heur is Associate Professor of Urban Studies at the Cosmopolis Centre for Urban Research, Vrije Universiteit Brussel, Brussels, Belgium.

David Bassens is Associate Professor of Economic Geography at the Cosmopolis Centre for Urban Research, Vrije Universiteit Brussel, Brussels, Belgium.

Disclosing Elite Ecologies

Methodologies For "Doing" Urban Elite Research

Edited by

Bas van Heur and David Bassens

Routledge
Taylor & Francis Group
LONDON AND NEW YORK

First published 2022
by Routledge
2 Park Square, Milton Park, Abingdon, Oxon OX14 4RN

and by Routledge
605 Third Avenue, New York, NY 10158

Routledge is an imprint of the Taylor & Francis Group, an informa business

British Library Cataloguing in Publication Data
A catalogue record for this book is available from the British Library

ISBN: 978-0-367-69675-7 (hbk)
ISBN: 978-0-367-69676-4 (pbk)
ISBN: 978-1-003-14281-2 (ebk)

Typeset in Minion Pro
by Newgen Publishing UK

Publisher's Note
The publisher accepts responsibility for any inconsistencies that may have arisen during the conversion of this book from journal articles to book chapters, namely the inclusion of journal terminology.

Disclaimer
Every effort has been made to contact copyright holders for their permission to reprint material in this book. The publishers would be grateful to hear from any copyright holder who is not here acknowledged and will undertake to rectify any errors or omissions in future editions of this book.

Contents

Notes on Contributors

Alberta Andreotti, Department of Sociology, University of Milan-Bicocca, Italy

David Bassens, Cosmopolis Centre for Urban Research, Vrije Universiteit Brussel, Brussels, Belgium.

Niall Cunningham, Geography, Durham University, UK.

Sarah Hall, School of Geography, University of Nottingham, Nottingham, United Kingdom.

Patrick Le Galès, Sciences-Po, CNRS Centre d'Etudes Européennes et de Politique Comparée and Sciences Po Urban School, Paris.

Lidia Katia Consiglia Manzo, Faculty of Sociology, The University of Trento, Trento, Italy.

Francisco Javier Moreno-Fuentes, Institute of Public Goods and Policies (IPP), Spanish National Research Council (CSIC), Madrid.

Bas van Heur, Cosmopolis Centre for Urban Research, Vrije Universiteit Brussel, Brussels, Belgium.

Fabio Vanin, Cosmopolis Centre for Urban Research at the Vrije Universiteit Brussel, Brussels, Belgium.

Lauren B. Wagner, Department of Technology and Society Studies, Maastricht University, Maastricht, the Netherlands.

Maëlys Waiengnier, Cosmopolis Centre for Urban Research, Vrije Universiteit Brussel, Brussels, Belgium.

An urban studies approach to elites: nurturing conceptual rigor and methodological pluralism

Bas van Heur and David Bassens

ABSTRACT

This special issue offers a set of methodologies to chart urban elites. Whereas most research has focused on the global super-rich, the authors in this special issue pay specific attention to the multidimensional urban geographies of elite reproduction and transformation, as elites depend on urban contexts for capital accumulation, consumption and leisure, and housing. In this article, we discuss theoretical and methodological antecedents in urban studies that have investigated economic elites. Building on but moving beyond these bodies of literature, we reject a-priori definitions of the size and shape of this social group and propose to pursue relational, place specific conceptualizations of elite composition and behavior. In particular, we argue, urban elite research would benefit from paying more attention to: (i) boundary work between elites and non-elites; (ii) intra-elite competition and distinction; (iii) national state spaces in determining elite composition; and (iv) the urban sense of belonging of economic elites.

Centering cities in elite research

In the recent crisis years, there has been increasing academic awareness of the elitist nature of the networks that govern our societies, polities, and economies (Mair, 2006; Streeck, 2006), and the growing social inequalities that result from that situation (Piketty, 2014). Fundamentally, such awareness has not been translated into social change to dislodge elites from the very power networks that lend them the capabilities to govern. While we can search for political-economic reasons for this "strange" fact (Bassens & van Meeteren, 2015; Crouch, 2011), this special issue aims to center the discussion on the more multidimensional urban geographies of elite reproduction and transformation, as urban elites depend to a significant extent on urban contexts for capital accumulation, consumption and leisure, housing, and other spheres of life. Researchers from various disciplines have started to recognize the relevant analytical and empirical lacunae, even though in geography this problematic of elite behavior and organization tends to be narrowed down to the question of the "super-rich" or "(ultra) high net worth" individuals (e.g., Beaverstock, Hall, & Wainwright, 2013; Beaverstock, Hubbard, & Short, 2004; Hay, 2013).

The limited attention (with some exceptions discussed in the next section) paid to cities and urban areas in elites research is striking, both as sites of research or as explanatory categories. This despite the fact that cities are a logical entry-point to charting elite geographies for reasons that are both agentic and structural in nature. Cities are places where administrative, political, cultural, and economic institutions cluster: institutions that lend elites power and legitimacy and vice versa. Cities are the "basing points" for capital (Friedmann, 1986), that is, the places where accumulated capital is (re)invested (Fernandez, Hofman, & Aalbers, 2016) and wealth reproduction is organized by a range of intermediaries (Beaverstock et al., 2013). Cities are replete with cultural resources that support distinct elite lifestyles (Bourdieu, 1984), and are marked by income and professional polarization, often in conjunction with patterns of socioeconomic segregation (Musterd, Marcińczak, van Ham, & Tammaru, 2017). Cities are the places where and from which elites govern formally and informally, both locally–as codified in growth coalition and urban regime thinking–and also on a global scale (Oosterlynck et al., 2018). Cities are spaces crosscut by globally mobile transitory populations, likely to produce a sociological layer on top of and often detached from local society, as evidenced from studies on expatriates' enclaves (Beaverstock, 2002). Therefore, why not start with cities when setting out to examine elite geographies? Beyond its rhetorical nature, this question is authentic in the sense that it touches on thorny conceptual and methodological issues when setting out to study urban elite geographies.

In fact, the overwhelming majority of academic and popular literature about elites tends to skate over their urban footprint. Instead, many studies focus on mapping income and wealth across countries. This applies both to the reports published by consultancy firms in finance and real estate on (ultra) high net worth individuals or billionaires (Capgemini, 2018; Knight Frank, 2018; Shorrocks, Davies, & Lluberas, 2017; Wealth-X, 2017, 2018) and to research by historians and economists on top incomes and "the one percent" (Alvaredo, Atkinson, Piketty, & Saez, 2013; Wisman, 2013). Although useful in tracing the increasingly unequal distribution of income and wealth, due to statistical data limitations the typical and sole level of analysis is the national level with no attention paid to the geographical distribution of income and wealth within countries. In addition, literature tells us very little about the extent to which this top layer of the population has any degree of power. As we know from elite theory and sociology, the prominence and dominance of wealthy people is not universal or to be taken for granted, as it depends on their particular relations to other types of elites (political, cultural, intellectual, etc.) and the wider population (Hartmann, 2004; Scott, 1996). Instead of adopting such global definitions and measurements of elites, we need to take a more contextual approach that investigates the place-specific evolution of the composition and practices of elites. Further, instead of analyzing economic elites as a social group that is relatively disconnected from other social groups, we need to understand the very power of these elites as the effect of their interactions (direct or indirect through all kinds of intermediaries and organizations) with other "non-elite" social groups. This special issue is concerned with offering a set of methodologies to chart elites in and through urban contexts.

Assembling theoretical and methodological antecedents in urban studies

In light of the state of affairs sketched out above, we are convinced that an interdisciplinary urban studies approach can offer viable methodological inroads toward dealing with the question of urban elite reproduction and transformation. This special issue builds on, but also moves beyond, at least three bodies of literature that have investigated economic elites in urban settings: (i) urban studies research into growth machines and urban regimes that focuses on the interactions between local political and business elites, (ii) research in economic geography on the "super-rich" and their housing and service needs, and (iii) ethnographic research that studies the social routines and internal workings of elite cultures.

While there are differences between the literature about growth machines (Logan & Molotch, 1987; Molotch, 1976; Rodgers, 2009) and about urban regimes (Imbroscio, 2003; Lauria, 1997), our main interest lies in their shared concern with "collaborative arrangements through which local governments and private actors assemble the capacity to govern" (Mossberger & Stoker, 2001, p. 812). This concern leads to a partial return to the so-called community power debate of the 1950s and 1960s in U.S. urban political sociology, and to a central problematic in elite theory. Whereas the "elitist school" in this debate argues that a core group of political and business elites determines the direction of city government (Hunter, 1953), the "pluralist school" rejects this view of a coherent urban elite and argues that we can only observe functional elites that exercise power over specific policy arenas and urban institutions (Dahl, 1961). One of the limitations of the literature on growth machines and urban regimes is that it assumes very strong social relationships between business and political elites, whereas many current city governments seem quite disconnected from business elites, with companies moving from one location to the next in search of better opportunities (see also Nevarez, 2002). This directs attention to the ways in which local political elites act in accordance with certain economic imaginaries (Jessop, 2004; Van Heur, 2010), even in the absence of strong social relationships between economic and political elites. This implies the need for a methodological openness to the position of different types of economic elites within specific cities, and their differential power relations to political or other elites (ranging from a strong coupling in line with urban regime and growth machine theory, to a complete decoupling). It also points to the need for a multiscalar understanding of urban elite power, since the "capacity to govern" cities is evidently shaped by discourses and networks that connect the urban to the national and supranational scales.

A second body of literature comprises more recent work on the super-rich in the field of economic geography (Beaverstock et al., 2004; Hay, 2013). This strand of literature by and large adopts the consultancy report definition of high net worth individuals to narrow down the empirical research on economic elites to those with at least a million U.S. dollars in financial assets. It can thus be subjected to the same criticism as discussed above, in that it uses generic, global definitions to describe socio-spatially variable processes. At the same time, this field has produced valuable empirical studies on super-rich mobilities (Birtchnell & Caletrío, 2013), housing (Butler & Lees, 2006), education (Waters, 2007), philanthropy (Hay & Muller, 2014), and services such as private wealth management (Beaverstock et al., 2013). Nevertheless, this body of work is limited in two ways. It is theoretically underdeveloped, as the analysis of the super-rich is largely concerned with elite social networks and elite spaces, and only to a very limited extent with the relationship between elite and non-elite groups, either directly or as mediated through all kinds of intermediaries. As a result, this strand of literature does not theoretically connect or contribute to the wider debate on social structure and inequality, and largely ignores the much more established tradition of elite sociology and theory in which these issues are repeatedly addressed. This neglect seems to be the result of the recurring argument that these elites are primarily global elites, only marginally connected to concrete urban spaces and societies. Literature about the super-rich is also underdeveloped due to its empirical selectivity. There is a tendency to focus on what we would describe as "glamorous cases" of elite distinction and behavior, such as neighborhoods of super-gentrification (Butler & Lees, 2006), elite enclave urbanism (Pow, 2011), private aviation (Budd & Hubbard, 2010), or luxury yachts (Spence, 2014). This ignores the wider range of elite practices that are perhaps less glamorous but most likely more common. Being attentive to the more mundane aspects of elite life in encounter with non-elites has also inspired the focus of the current special issue.

The third research strand relevant to this special issue is anthropological work on elites as well as, more broadly, research into elites that adopts substantive ethnographic approaches. Some of the work in the other two bodies of literature discussed above includes such approaches as part of a wider social scientific methodological repertoire, but it usually remains a minor part that is instrumentally mobilized in the service of a primarily political economic explanation of elite rule. By contrast, ethnographies of elites contribute more detailed and contextualized descriptions to the analysis, which pay particular attention "to the internal workings of elite cultures, to the way they construct, employ, and perpetuate their power and influence, and to the social perception and cultural valuation of elites among the larger public" (Salverda & Abbink, 2012, p. 3). Bourdieu's work on social distinction and habitus formation (e.g., Bourdieu, 1984, 1989) has been very influential in this body of work, but the specific interest in elite (self-)representation, symbolism, lifestyles, behaviors, and the ways in which (and to what extent) these contribute to elite legitimacy and rule is shared by many authors, both old and new. This includes Veblen (1899) on elite taste and conspicuous consumption, Mills (1956) on metropolitan clubs, Khan (2011) on elite boarding schools, and Pinçon and Pinçon-Charlot (2013) on the lifestyles of the French upper bourgeoisie, to mention just a few. Almost all of this research is strongly case study driven. This of course is its main strength as well as its weakness, because it becomes difficult to generalize from the various case studies that above all show the extraordinary socio-spatial variation of elite cultures. The current special issue adopts the inductive sensibility to local practices and representations–against the strong claim that we are moving toward some kind of "transnational capitalist class" (Sklair, 2001)–but integrates this into a multiple method framework that allows for comparisons between cases and, on this basis, theorization about urban elites that transcends the single case. In line with Daloz (2010, 2013)), the aim is not to produce another grand theory about elites, but a revaluing of urban elites from a comparative perspective that avoids the pitfalls of ethnocentrism and reductionism.

Toward relational conceptualizations of elite composition and behavior

A multidimensional and expanded reading of elites

What binds together these different approaches to economic elites is a rejection of a-priori definitions of the size and shape of this social group, instead proposing more relational, contextual understandings of the composition and behavior of elites: who they are, what they do, the relations they develop with other social groups, and the power they exercise, is relatively place specific. This comes at the price of a more global descriptive definition, but it has the advantage of potentially explaining elite power. It allows us to analyze and understand better *how* economic elites–labeled in overlapping but slightly different literatures, as the super-rich (Beaverstock et al., 2004; Hay, 2013), the one percent (Keister & Lee, 2014), high net worth individuals (Mason & Harrison, 2015), and billionaires (Neumayer, 2004)–exercise power over other non-elite groups by commanding a disproportionate share of global income and wealth. This involves much more than merely privileged access to economic resources, since the exercise of power is unavoidably intermeshed with often very particular relations to other actors, organizations, and institutions. These relations and settings deserve empirical investigation, not theoretical speculation as to the ontological centrality of economic elites in the current era. In undertaking these empirical investigations in particular cities, it is not unlikely that we will regularly observe economic elites dependent on (not independent from) a complex mix of "old" and "new" money, family relations, educational trajectories, local or national patriotism and other types of civic sentiment, and the local political structure, as well as the extent to which non-elite groups contest or follow elite practices.

Various contributors to this special issue emphasize the ways in which definitional and operational discussions concerning the concept and category of elites are intertwined with substantive reflections on the position of those elites within the wider socio-spatial structure. Niall Cunningham (2017) discusses the ways in which the Bourdieusian analysis used in the BBC-commissioned Great British Class Survey involves a conceptualization of a "new ordinary elite" that comprises seven percent of the population in contrast to "the one percent" super-rich that have received most attention in urban and economic geography. This more inclusive definition is a direct result of the adopted multidimensional capitals approach, in which elite status is derived not only from the mobilization of economic, but also cultural and social resources, which sets these elites apart from the population at large. This has clear spatial consequences: whereas literature about the super-rich tends to focus on a select number of exclusive neighborhoods characterized by gated communities, absentee ownership, and public disengagement (Atkinson, 2016), Cunningham points to elite concentrations in a wider range of neighborhoods; areas in which high levels of economic capital are combined with strong social networks and cultural engagement. In their article about the local "rootedness" of managers in Madrid, Milan, Lyon, and Paris, Alberta Andreotti, Le Galès, and Moreno-Fuentes (2018) also explicitly reject a too narrow concern with a super-rich global elite and argue that the prime focus should instead be on the upper middle class, a group they see located at the top five to fifteen percent of the income hierarchy. In line with Cunningham, this leads them to conclude that members of this class continue to invest economic, social, and cultural resources in their city and neighborhoods (often building on established family and social ties that span generations), and only use "urban exit" strategies in a select number of contexts.

Boundary work between elites and non-elites

The adoption of a relational as well as a contextual approach toward investigating elites, and the acknowledgement that elites actively participate in the messiness of urban life, unavoidably directs attention to the unequal power relations between elite and non-elite social groups. Surprisingly, this dimension is rarely picked up on in current debates within the field of urban geography, whereas it is central to the tradition of sociological research on elites. As already mentioned above, this includes Veblen (1899) on conspicuous consumption, and Bourdieu (1984, 1989) on social distinction and habitus formation, but also Max Weber on status, Georg Simmel on the anonymity of metropolitan life and self-presentation in the face of strangers, Erving Goffman on social interaction and impression management, as well as others (for a concise overview of these and other sociological approaches, see Daloz, 2010). Critically engaging with the work of Bourdieu, Michèle Lamont (1992; Lamont & Fournier, 1992) has made important contributions to the study of symbolic boundaries and inequality. Along similar lines but somewhat distinct from this tradition of North American sociology, cultural sociologists in the U.K. have pursued more spatialized and "cultural" versions of class analysis by investigating the roles of lifestyles and cultural boundaries, pointing to an "elite habitus" (Savage, 2017) and the limited exposure of elites to urban public settings (Atkinson, 2016).

The current special issue points to the importance of re-engaging with this tradition. Having argued that the upper middle class in European cities is rooted, Andreotti and colleagues then go on to show how the managers they investigated engage in a game of distance and proximity in order to control their relations with the non-elites living in their neighborhood. This management of differences is less centered on spatial segregation and more on controlling interaction within social networks, and in the context of public services and settings. In her ethnographic study on the gentrified Brooklyn Park Slope neighborhood, Manzo (2017) provides a counter example in that she investigates two critical cases in which elites are unlikely to be ruling actors: a martial arts studio and a food cooperative market. Both being places that are characterized by collective or egalitarian social codes and with a history of lower-class membership. As Manzo argues, in these settings, where elites are unable to control the dominant values and norms, they resort to more specific micro-strategies in order to still distinguish themselves from non-elites: ranging from reluctance to take lessons from karate practitioners with a higher belt color but a lower socioeconomic status, to the tendency to volunteer for the symbolically more "powerful" positions of group coordinator instead of food packager in the food coop. Similarly, and not surprisingly also based on in-depth ethnographic research, Wagner (2019) dwells on this out-of-place character of elites as a creative tension to understand elite reproduction. Her contribution shows how the European-Moroccan diaspora upon returning to Fez "accidentally" receive elite status vis-à-vis the non-diasporic Moroccans, through their leisurely consumption practices. Lastly, Fabio Vanin (2019) points to the key importance of fear and the perception of security in mediating the relations between elites and non-elites. Drawing on case study work in Porto Alegre–one of the most polarized cities in Brazil and with high crime rates–Vanin shows how the security infrastructure around gated communities contributes to the production of a landscape of fear in which the interactions between elites and non-elites are minimized and highly controlled. At the same time, the gated community and the ever-increasing sophistication of security infrastructures act as tools of distinction and as aspirational models for the upper middle class.

Intra-elite competition and distinction

This last observation directs attention toward thedivisions and distinctions between elites. Quite clearly, as Jonathan Beaverstock is fond of saying, there are significant wealth inequalities among the super-rich (Beaverstock & Faulconbridge, 2014). We follow Chrystia Freeland's observation (2012) that the biggest threat to the rule of the super-rich might come not from non-elite social groups, but from other economic elites that feel threatened by and/or want to mimic the lifestyles of the super-rich. Moving beyond the too narrow focus on the super-rich, this can be stated in more general terms, as it points to the need to differentiate between elites and to acknowledge that different elites do not necessarily share the same ways of life or the same interests (Salverda & Abbink, 2012). It further complicates simple assumptions concerning elite rule and brings us back to the traditional literature about elite studies and its identification of functional elites whose power is limited to specific policy arenas and urban institutions. One of the major consequences of this differentiation of elites is that we can observe strong intra-elite competition and distinction, with elites competing between each other for social dominance. In line with Daloz (2010, 2013)), it is likely that intra-elite competition and distinction is of more importance to elites than dominance over the lower strata, since this is largely taken for granted. This raises important questions concerning the relations and conflicts between, for example, the "new" super-rich and the established economic elites, and the ways in which the struggle around elite positional goods can be understood as a key driving force regarding urban change.

In this special issue, intra-elite distinction and competition are perhaps most apparent in the contribution by Sarah Hall (2018). Her article on the Chinese financial elites in London's financial district builds on established work in the field of economic geography about the international mobility of financial elites and high-end labor market migration, but directs attention toward state-orchestrated Chinese migration to the United Kingdom and the role of nationality and ethnicity in shaping competition within London-based financial elite networks. As Hall shows, due to limited ethnic diversity within the financial labor market, Chinese migrants experience various types of challenges when trying to convert their educational credentials into a career with international banks. The contribution by David Bassens, Bas van Heur, and Maëlys Waiengnier (2018), dealing with cultural patronage and urban elites in Brussels, tackles intra-elite distinction from a different angle. Drawing on cultural patronage membership lists from five "high culture" institutions, their analysis hints at various moments of intra-elite distinction. Each cultural institution tends to cater to its own elite social circles, with very limited membership overlap between the institutions. In addition, the sampled elites are largely Belgian nationals, suggesting that non-Belgian elites in Brussels–such as those working for European institutions or the lobby complex–frequent different organizations. Further, although elites live in the same set of neighborhoods, those working in the arts world are concentrated more than other elites in the gentrifying areas of Brussels, hinting at their role as bridging intermediaries between established elites and new cultural scenes characterized by "emerging cultural capital" (Savage, Hanquinet, Cunningham, & Hjellbrekke, 2018).

From global, footloose elites to locally rooted elites? state spaces and urban belonging

Such a contextual and relational approach toward conceptualizing and researching elites quasi-automatically leads to a critique of literature that understands elites as fundamentally globalized and footloose. Too often the focus has been solely on very narrow networks of global elites linked to financial and professional services (e.g., Beaverstock, 2002; Hall, 2009), which usually tends to underplay the relevance and power of more parochial elites and their connections in urban settings (Andreotti, Le Galès, & Moreno-Fuentes, 2015). This is a debate that becomes all the more relevant when discussing second (or third) tier cities in a variety of regional and national contexts, which are not necessarily well connected to global accumulation processes (Van Heur & Lorentzen, 2012). Our special issue contributes to mapping elites in this larger variety of urban contexts by offering studies on Brussels, Fez, London, Lyon, Madrid, Manchester, Milan, New York City, Paris, and Porto Alegre. While there is an obvious gravitational pull toward mainstay global cities, second-tier cities, as well as cities in emerging economies, are also sites of study. The variety of cases should allow the reader to tease out the similarities and differences between various urban contexts.

Two preliminary observations can be made in this special issue introduction. First, whereas most of the literature about economic elites and in particular the super-rich has prioritized the analysis of financial elites and the ecosystem of companies within which these elites mostly circulate, contributors to this special issue point to the continuing importance of the national context and the role of the state in structuring the composition and practices of urban elites. Andreotti and colleagues justify their empirical focus on upper middle class managers by pointing to the importance of urbanized *bourgeoisies* in the development of European cities, and their breakdown into four key social groups (in addition to managers, also entrepreneurs, high-ranking bureaucrats, and old and new professionals) with relevant differences between European countries. Similarly, Bassens and colleagues highlight the national character of elites in Brussels, with the negligible presence of non-Belgian elites and a remarkably strong presence of nobility in the investigated sample, pointing to a reproduction of elites from the *Ancien Régime* until today. On the other side of the world, Vanin points to a growing distrust–after the end of the military dictatorship in 1985–of public authorities among Brazilian elites and how this has shaped their security-oriented and fear-ridden lifestyles. For London, perhaps the global city *par excellence*, Cunningham offers a particularly British perspective on the geography of elites, understanding the composition of the elites in London and other urban areas as part of a broader transformation of British state space. For the Moroccan case presented by Wagner, the enthusiasm with which the state evaluates the seasonal return of the diaspora enables their "accidental" elite formation, reminiscent of sending-state strategies in other countries with large diaspora populations (Ho, 2011; Krijnen, Bassens, & van Meeteren, 2017).

Second, the elites examined in this special issue are quite obviously mobile, but also clearly rooted. Andreotti and colleagues explain this local rootedness on the basis of active local family ties and inheritance strategies. Bassens and colleagues point to strong continuities in the residential pattern of Brussels elites between the nineteenth century and now, which might also be partially explained as a result of these local ties and inheritances. Andreotti and colleagues also hint at a role for elective (not just inherited) belonging (Savage, Bagnall, & Longhurst, 2005), which is developed in greater detail by Cunningham, who explicitly identifies the mapped elite neighborhoods as areas characterized by self-selection, shared cultural tastes, and active identification. The dark side of elective belonging and the closure of elite networks becomes visible in the contribution of Hall, in which she argues that the limited ethnic diversity within London's financial labor market curtails work opportunities for Chinese financial elites and leads at least some to consider London as only a temporary stepping stone on the way to a financial center such as Beijing or Singapore that is closer to home.

Conceptual rigor, methodological pluralism

In conclusion, we are pleased that *Urban Geography* has offered us the opportunity to edit this special issue, since it has always adopted "radical openness as method" (see The Editors, 2015), which stimulates rigorous yet multidimensional engagements across disciplines. In line with the editorial position of the journal, we believe this requires a style of theory development that is contextual (Cartier, in The Editors, 2015) as well as substantively interdisciplinary, drawing on methods and theories from inside and outside of urban geography (Cooke & Ward, in The Editors, 2015). This special issue with its interdisciplinary urban studies character has been put together in precisely that spirit: taking inspiration from urban and economic geography, elite theory and urban sociology, cultural sociology, political economy, anthropology, criminology, architecture, and migration studies. Such an interdisciplinary approach also has methodological motivations and consequences. The special issue provides readers with various empirical inroads into the study of urban elites, drawing on a related variety of disciplinary backgrounds, and opening up the opportunity for methodological cross-fertilization.

On the basis of these methodologically reflexive accounts, the reader should become aware of various original ways of engaging in urban elite research, which can then be reproduced in other urban settings. Many of these methods offer and reflect on established, but also innovative approaches for "doing urban elite research" from a range of disciplines. In the process, a range of new sources of data are presented and discussed in terms of their explanatory relevance with regard to unveiling elite networks and spaces. Hall uses a new dataset of national insurance registrations to provide a detailed mapping of Chinese migrants in London, and draws on social media data to identify and gain access to interviewees. Bassens and colleagues rely on cultural patronage lists, social media, and accessible online data to identify the occupations and residential addresses of urban elites in Brussels, cross-checked by the visual documentation of elite housing. In addition to interviews and video/photo surveys, Vanin conducts detailed visual and morphological analyses of elite housing in Porto Alegre. Both Wagner and Manzo engage in deep ethnographies of "mundane" places in respectively Fez and Brooklyn not commonly associated with urban elites (food coop, boxing club, tourist nightlife). Andreotti and colleagues offer an advanced comparative setup with multiple empirical entry points to gain access to and construct their sample of elites. Lastly, Cunningham reflects on how the skewed and non-representative nature of the data produced by the self-selected Great British Class Survey is not a weakness, but actually a possible strength in conducting research on urban elites. It is our conviction that the set of papers presented here can serve as a resourceful toolbox for urban geographers that aim to engage critically and reflexively with a range of methodologies to carry out urban elite research. In our view, urban elite geographies, with their global *and* contextual character, are a key theme around which such a practice of engaged pluralism can emerge, much to the benefit of those interested in productively switching research frames (see Van Meeteren, Bassens, & Derudder, 2016). Accordingly, it is our aim and hope that the special issue will lure elite researchers with different disciplinary approaches into the field of urban studies as a meaningful zone of intellectual exchange.

Disclosure statement

No potential conflict of interest was reported by the authors.

References

Alvaredo, Facundo, Atkinson, Anthony B., Piketty, Thomas, & Saez, Emmanuel. (2013). The top 1 percent in international and historical perspective. *Journal of Economic Perspectives*, *27*(3), 3–20.

Andreotti, Alberta, Le Galès, Patrick, & Moreno-Fuentes, Francisco Javier. (2015). *Globalised minds, roots in the city: Urban upper-middle classes in Europe*. Hoboken N.J: Wiley-Blackwell.

Andreotti, Alberta, Le Galès, Patrick, & Moreno-Fuentes, Francisco Javier. (2018). The challenge of researching "partial exit" and "rootedness" among upper-middle classes in European cities. *Urban Geography*. doi:10.1080/02723638.2018.1472443

Atkinson, Rowland D. (2016). Limited exposure: Social concealment, mobility and engagement with public space by the super-rich in London. *Environment and Planning A*, *48*(7), 1302–1317.

Bassens, David, van Heur, Bas, & Waiengnier, Maëlys. (2018). Follow the money: Cultural patronage and urban elite geographies. *Urban Geography*, 1–28. doi:10.1080/02723638.2018.1449429

Bassens, David, & van Meeteren, Michiel. (2015). World cities under conditions of financialized globalization: Towards and augmented world city hypothesis. *Progress in Human Geography*, *39*(6), 752–775.

Beaverstock, Jonathan V. (2002). Transnational elites in global cities: British expatriates in Singapore's financial district. *Geoforum*, *33*(4), 525–538.

Beaverstock, Jonathan V., & Faulconbridge, James. (2014). Wealth segmentation and the mobilities of the super-rich: A conceptual framework. In Thomas Birtchnell & Javier Caletrío (Eds.), *Elite Mobilities* (pp. 40–61). (2013). London: Routledge.

Beaverstock, Jonathan V., Hall, Sarah, & Wainwright, Thomas. (2013). Servicing the super-rich: New financial elites and the rise of the private wealth management retail ecology. *Regional Studies*, *47*(6), 834–849.

Beaverstock, Jonathan V, Hubbard, Philip, & Short, John Rennie. (2004). Getting away with it? Exposing the geographies of the super-rich. *Geoforum*, *35*(4), 401–407.

Birtchnell, Thomas, & Caletrío, Javier (Eds.). (2013). *Elite Mobilities*. London: Routledge.

Bourdieu, Pierre. (1984). *Distinction: A social critique of the judgement of taste.* London: Routledge.
Bourdieu, Pierre. (1989). Social space and symbolic power. *Sociological Theory, 7*(1), 14–25.
Budd, Lucy C.S., & Hubbard, Phil. (2010). The 'bizjet set': Business aviation and the social geographies of private flight. In Jonathan V Beaverstock, Ben Derudder, James Faulconbridge, & Frank Witlox (Eds.), *International business travel in the global economy* (pp. 85–104). London: Ashgate.
Butler, Tim, & Lees, Loretta. (2006). Super-gentrification in Barnsbury, London: Globalization and gentrifying global elites at the neighbourhood level. *Transactions of the Institute of British Geographers, 31*(4), 467–487.
Capgemini (2018). *World Wealth Report 2018*, Paris: Author.
Crouch, Colin. (2011). *The strange non-death of neoliberalism.* Cambridge, U.K.. Polity Press.
Cunningham, Niall. (2017). Making and mapping Britain's "new ordinary elite". *Urban Geography.* doi:10.1080/02723638.2017.1390721
Dahl, Robert A. (1961). *Who governs? democracy and power in an American city.* New Haven: Yale UP.
Daloz, Jean-Pascal. (2010). *The sociology of elite distinction: From theoretical to comparative perspectives.* Basingstoke: Palgrave Macmillan.
Daloz, Jean-Pascal. (2013). *Rethinking social distinction.* Basingstoke: Palgrave Macmillan.
Fernandez, Rodrigo, Hofman, Annelore, & Aalbers, Manuel. (2016). London and New York as a safe deposit box for the transnational wealth elite. *Environment and Planning A, 48*(12), 2443–2461.
Frank, Knight. (2018). *The Wealth Report 2018.* London: Knight Frank.
Freeland, Chrystia. (2012). *Plutocrats: The rise of the new global super-rich and the fall of everyone else.* New York: Penguin Press.
Friedmann, John. (1986). The world-city hypothesis. *Development and Change, 17*(1), 69–83.
Hall, Sarah. (2009). Financialised Elites and the changing nature of finance capitalism: Investment bankers in London's financial district. *Competition & Change, 13*(2), 173–189.
Hall, Sarah. (2018). Reframing labour market mobility in global finance: Chinese elites in London's financial district. *Urban Geography.* doi:10.1080/02723638.2018.1472442
Hartmann, Michael. (2004). *Elitesoziologie: Eine Einführung.* Frankfurt: Campus Verlag.
Hay, Iain. (2013). *Geographies of the super-rich.* Cheltenham: Edward Elgar.
Hay, Iain, & Muller, Samantha. (2014). Questioning generosity in the golden age of philanthropy: Towards critical geographies of super-philanthropy. *Progress in Human Geography, 38*(5), 635–653.
Ho, Elaine Lynn-Ee. (2011). "Claiming" the diaspora: Elite mobility, sending state strategies and the spatialities of citizenship. *Progress in Human Geography, 35*(6), 757–772.
Hunter, Floyd. (1953). *Community Power Structure.* Chapel Hill: University of North Carolina Press.
Imbroscio, David L. (2003). Overcoming the neglect of economics in urban regime theory. *Journal of Urban Affairs, 25*(3), 271–284.
Jessop, Bob. (2004). Cultural political economy, the knowledge-based economy, and the state. In Andrew Barry & Don Slater (Eds.), *The technological economy* (pp. pp. 144–146). London: Routledge.
Keister, Lisa A., & Lee, Hang Young. (2014). The one percent: Top incomes and wealth in sociological research (2014). *Social Currents, 1*(1), 13–24.
Khan, Shamus Rahman. (2011). *Privilege: The making of an adolescent Elite at St. Paul's boarding school.* Princeton: Princeton UP.
Krijnen, Marieke, Bassens, David, & van Meeteren, Michiel. (2017). Manning circuits of value: Lebanese professionals and expatriate world-city formation in Beirut. *Environment and Planning A, 49*(12), 2878–2896.
Lamont, Michèle. (1992). *Money, morals, and manners: The culture of the French and the American upper-middle class.* Chicago: University of Chicago Press.
Lamont, Michèle, & Fournier, Marcel (Eds). (1992). *Cultivating differences: Symbolic boundaries and the making of inequality.* Chicago: University of Chicago Press.

Lauria, Mickey (Ed). (1997). *Reconstructing urban regime theory: Regulating urban politics in a global economy*. London: Sage.

Logan, John, & Molotch, Harvey. (1987). *Urban fortunes: The political economy of place*. Berkeley: University of California Press.

Mair, Peter. (2006). Ruling the void: The hollowing of western democracy. *New Left Review, 42* (Nov–Dec), 25–51.

Manzo, Lidia Consiglia Katia. (2017). Naked elites: Unveiling embodied markers of superiority through co-performance ethnography in gentrified Brooklyn's Park Slope. *Urban Geography*, 1–20. doi:10.1080/02723638.2017.1381535.

Mason, Colin M., & Harrison, Richard T. (2015). Business angel investment activity in the financial crisis: UK evidence and policy implications. *Environment and Planning C, 33*(1), 43–60.

Mills, C. Wright. (1956). *The Power Elite*. New York: Oxford UP.

Molotch, Harvey. (1976). The city as growth machine. *American Journal of Sociology, 82*(2), 309–332.

Mossberger, Karen, & Stoker, Gerry. (2001). The evolution of urban regime theory: The challenge of conceptualization. *Urban Affairs Review, 36*(6), 810–835.

Musterd, Sako, Marcińczak, Szymon, van Ham, Maarten, & Tammaru, Tiit. (2017). Socioeconomic segregation in European capital cities: Increasing separation between poor and rich. *Urban Geography, 38*(7), 1062–1083.

Neumayer, Eric. (2004). The super-rich in global perspective: A quantitative analysis of the Forbes list of billionaires. *Applied Economics Letters, 11*(13), 793–796.

Nevarez, Leonard. (2002). Efficacy or legitimacy of community power? A reassessment of corporate elites in urban studies. In John Eade & Christopher Mele (Eds.), *Understanding the City: Contemporary and future perspectives* (pp. 379–396). Oxford: Blackwell.

Oosterlynck, Stijn, Beeckmans, Luce, Bassens, David, Derudder, Ben, Segaert, Barbara, & Braeckmans, Luc. (2018). *The City as a Global Political Actor*. London and New York: Routledge.

Piketty, Thomas. (2014). *Capital in the Twenty-First Century*. Cambridge (U.K: Harvard University Press.

Pinçon, Michel, & Pinçon-Charlot, Monique. (2013). *La violence des riches: Chronique d'une immense casse sociale*. Paris: La Découverte.

Pow, Choon-Piew. (2011). Living it up: Super-rich enclave and transnational elite urbanism in Singapore. *Geoforum, 42*(3), 382–393.

Rodgers, Scott. (2009). Urban growth machine. In Rob Kitchin & Nigel Thrift (Eds.), *International encyclopedia of human geography* (pp. pp. 40–45). Amsterdam: Elsevier.

Salverda, Tijo, & Abbink, Jon. (2012). Introduction: An anthropological perspective on elite power and the cultural politics of elites. In Jon Abbink & Tijo Salverda (Eds.), *The anthropology of elites: Power, culture, and the complexities of distinction* (pp. 1–28). New York: Palgrave Macmillan.

Savage, Mike. (2017). The elite habitus in cities of accumulation. In Suzanne Hall & Ricky Burdett (Eds.), *The SAGE Handbook of the 21st Century City* (pp. 71–86). London: SAGE Publications.

Savage, Mike, Bagnall, Gaynor, & Longhurst, Brian J. (2005). *Globalization and Belonging*. London: Sage Publications.

Savage, Mike, Hanquinet, Laurie, Cunningham, Niall, & Hjellbrekke, Johs. (2018). Emerging cultural capital in the city: Profiling London and Brussels. *International Journal of Urban and Regional Research, 42*(1), 138–149.

Scott, John. (1996). *Stratification & power: Structures of class, status and command*. Cambridge: Polity Press.

Shorrocks, Anthony, Davies, Jim, & Lluberas, Rodrigo. (2017). *Credit suisse global wealth report 2017*. Zurich: Credit Suisse AG.

Sklair, Leslie. (2001). *The transnational capitalist class*. Oxford: Wiley-Blackwell.

Spence, Emma. (2014). Towards a more-than-sea geography: Exploring the relational geographies of superrich mobility between sea, superyacht and shore in the Cote d'Azur. *Area, 46*(2), 203–209.

Streeck, Wolfgang. (2006). The study of interest groups: Before 'the Century' and After. In C. Crouch & W. Streeck (Eds.), *The diversity of democracy: Corporatism, social order and political conflict*. Cheltenham: Edward Elgar.

The Editors. (2015). New urban geographies. *Urban Geography, 36*(3), 327–339.

Van Heur, Bas. (2010). Beyond regulation: Towards a cultural political economy of complexity and emergence. *New Political Economy, 15*(3), 421–444.

Van Heur, Bas, & Lorentzen, Anne (Eds). (2012). *Cultural political economy of small cities*. London and New York: Routledge.

Van Meeteren, Michiel, Bassens, David, & Derudder, Ben. (2016). Doing global urban studies: On the need for engaged pluralism, frame switching, and methodological cross-fertilization. *Dialogues in Human Geography, 6*(3), 296–301.

Vanin, Fabio. (2019). Investigating urban elites through fear and security in Porto Alegre. *Urban Geography, 40*(5), 732–746.

Veblen, Thorstein. (1899). *The theory of the leisure class: An economic study of institutions*. London: Macmillan.

Wagner, Lauren. (2019). Accidental elites: Accumulating effects of diasporic encounters. *Urban Geography, 40*(5), 666–684.

Waters, Johanna L. (2007). 'Roundabout routes and sanctuary schools': The role of situated educational practices and habitus in the creation of transnational professionals. *Global Networks, 7*(4), 477–497.

Wealth-X. (2017). *World ultra wealth report 2017*. New York: Author.

Wealth-X. (2018). *Billionaire census 2018*. New York: Author.

Wisman, Jon D. (2013). Wage stagnation, rising inequality and the financial crisis of 2008. *Cambridge Journal of Economics, 37*(4), 921–945.

Making and mapping Britain's "new ordinary elite"

Niall Cunningham

ABSTRACT

There has been a sharp intensification in public and academic interest in differing conceptions of an urban "elite" in recent times. However, the concept of social class in the construction and reproduction of urban elites has remained either an implied or unexplored concern. The purpose of this paper is to explore the empirical manifestations and methodological issues surrounding the definition of an elite which arose from the BBC's Great British Class Survey experiment. This paper builds on our commonplace understanding of an elite as economically distinct by focussing on their social resources and patterns of cultural consumption, based upon a Bourdieusian"capitals" approach to social class, and highlighting dimensions of this cadre which have hitherto received scant attention in recent public and academic debate.

Introduction

In the first years of the twenty-first century there has been both growing interest and disquiet over the increasing inequalities which characterise so many global and national economies. This has led to a burgeoning of research in the area and the effects of the concentration of wealth as an inherent characteristic of capitalism (Piketty, 2014) or on the negative social consequences of such inequality (Wilkinson & Pickett, 2009). What has been revealing about much of this work has been the absence or elision of social class as a means for interpreting and understanding the manifestation, reproduction and intensification of these processes of social, economic and spatial inequality. This paper acts as part of the ongoing process of reinstating and spatializing social class as essential to our understanding of contemporary patterns of inequality (Parker, Uprichard, & Burrows, 2007). For Parker and his colleagues, as for others (Butler & Lees, 2006; Butler with Robson, 2003; Savage, Bagnall, & Longhurst, 2005; Slater, 2009), we must make residential structures of social class a core component of our analyses. They are not simply subsidiary to or merely reflective of class position, they are also generative in the construction of classed identities. Furthermore, for Parker et al. (2007) the need to spatialize class gains urgency from the challenges and opportunities posed by geodemographics and the rise of what they term "commercial sociology" being conducted in the private sector but having an increasing impact upon the academic analysis of urban inequality (Atkinson, 2015; Atkinson, Parker, & Burrows, 2017;

Webber, 2007; Webber & Burrows, 2015). Due in part to reasons of academic hubris and self interest, such developments have not, argue Savage and Burrows (2007) been awarded the serious attention they deserve, posing a challenge of existential proportions to the social sciences and the discipline of sociology in particular It is against this intellectual backdrop which the BBC's *Great British Class Survey* (GBCS) experiment and its particular conception of the *Elite* can be seen to have emerged. However, it is very important to avoid the sort of "presentist" mentality which critics such as Robert Dodgshon (2008) have argued has infected contemporary geographical discourse. The notion of the "spatialisation of class" can clearly be traced back to the 1960s and the pioneering work of scholars such as Ray Pahl, Rex and Moore (1967) at the class/race interface and a specific concern with the role of housing type in the residualisation or reinforcement of class position at the bottom of the social spectrum, something that has remained a concern of social scientists working across disciplinary boundaries (Brown & Cunningham, 2016; Burrows, 1999; Forrest & Murie, 1983; Power & Tunstall, 1995; Savage, 2010).

The specific purpose of this paper is to explore some of the methodological issues and implications of an attempt to reassert the centrality of social class in our understanding of social inequality through the development of a new social class schema and the articulation of a new and rather more extensive conceptualisation of an "elite" than is generally recognised amongst geographers, sociologists and the public at large. According to the *GBCS* "new class model" developed by Savage et al. (2013), the *Elite* comprise seven percent of the population and is therefore a significant component of society rather than a tiny cadre of super-rich individuals who have provided the focus for other researcher in the field of urban elites (Atkinson, 2015; Beaverstock, Hubbard, & Scott, 2004; Butler & Lees, 2006; Webber & Burrows, 2015). Indeed, it has been defined by Savage et al. (2015) as a "new, ordinary" *Elite*. The empirical data for the construction of this new *Elite* came from the BBC's *GBCS* experiment, a project which has opened up a new debate on the contemporary meaning of social class and its relationship to inequality which has extended well beyond the usual academic discourse (2015; Savage et al., 2013). The theoretical basis for the construction of this project was heavily informed by the French sociologist Pierre Bourdieu, and specifically his concept of social class as based upon one's stocks of social, economic and cultural capital (1984). Thus, this configuration shifts the focus solely from economic or occupational position as the basis for social class position and it is the focus on cultural capital, which builds upon a established and growing school of "cultural class analysis" (Atkinson, 2010) which has been perhaps the most contentious element. Thus, this re-imagining of social class has been strongly contested on both methodological and substantive lines (Dorling, 2013; Mills, 2014; Mills, 2015; Savage et al, 2015a; Skeggs, 2015), criticisms which this paper will engage with. However, another purpose of this paper is to focus on the spatial and temporal rhythms of participation in the survey as an instrument for socio-spatial research and the light this may shed on a particular conception of an urban elite as a stratified and highly self-propagating entity. Contrary to the recent work of Atkinson (2015) on an albeit more exclusive definition of an urban elite, this more accessible vision sees this group as egregiously revealing of their social status and habits. However, whilst displaying distinctive patterns of engagement and spatial settlement, this *Elite* is far from amorphous, displaying differential stocks of economic, cultural and social capital across space. The paper will also address this *Elite* class in relation terms by drawing attention to their economic resources in contrast to the wider populations of Britain's largest urban centres drawing on representative data.

Context: inequality, the "death" of social class and the genesis of the elite

It might be supposed that the concept of social class would be central to an understanding of inequality, the broader issue within which any consideration or conceptualisation of elites must be framed. However, since the 1980s social class had fallen out of favour with academics, politicians and the public at large. The reasons for this are complex. From the early 1980s, the "cultural turn" within the social sciences was leading academics away from over-arching structuralist explanations for inequality towards a firmer consideration of the place of the individual and the way in the experience of class was mediated through different constituent identities such as race, gender and sexuality. Ray and Sayer (1999, pp. 13–14) summarise this intellectual shift neatly as the transition "from economy to culture" which accrued from the "realization that most non-class social divisions could not be explained on the model of class". In this way, Ray and Sayer were articulating part of a broader ontological concern on the part of social scientists about the difference between ascriptive and subjective social and cultural differences. These were intellectual developments not solely limited to the British context; Jan Pakulski (1993), along with Pakulski and Walters (1996) pointed to transnational processes of de-industrialisation, the fragmentation of identities and appropriated Benedict Anderson's (1991) theorisation of nationalism for a critique of the cultural turn in the emergence of new "imagined communities". In the US, David Grusky and Kim Weeden's empirical analyses Weeden and Grusky (2005) led them to argue that the "big classes" employed by sociologists had little meaning in the real world, and argued instead for an analysis of "micro-classes", effectively much smaller occupational groups, which they argued constituted much stronger and more realistic components of a class structure. So whilst still utilising occupation as the basis of understanding social structure, the evidence Weeden and Grusky (2012) present suggests that these micro-classes are more powerful the big classes of established sociology in predicting a range of social, cultural and attitudinal outcomes.

In the UK, the discipline of sociology was riven by increasing acrimonious internal disagreement over the role and relevance of social class, with Ray Pahl arguing in 1989 (p. 710) that "class as a concept was ceasing to do any useful work for sociology". Such a claim came against an economic setting of industrial decline, rapid financialisation of the economy and a breakdown in traditional class identities. The rampant inequalities now firmly associated with the UK and many other developed economies can be directly associated with the de-regulation of the financial services sector in the 1980s (Tanndal & Waldenström, 2016). In the UK context we can add to that mix the poisonous politics of class warfare evident in the Thatcher government's handling of the Miners' Strike of 1984–85 and the aggressive counterexpropriation of public sector housing through the "right to buy" scheme and we can start to see the political mechanisms by which social class had become politically neutered by the 1990s. So politically moribund was social class as a concept that consecutive opposing Conservative and Labour Prime Ministers could agree that it was either dead or dying. For the Conservative John Major such a claim was presented merely in aspirational terms, striving in 1990, as he put it, for "a classless society" (Turner, 2013, pp. 1–2) His Labour successor Tony Blair was somewhat more robust, proclaiming just seven years later that "the class war is over" (BBC, 1999). At the start of the new millennium and against the rise of the internet and consumer culture, the eclipse if not the actual death of class, appeared confirmed. However, it is interesting to note that even in the thinking of Beck and Beck-Gernsheim (2002, p. 203) this death was incomplete as he argued that class had become a "zombie category", "dead but still alive". So class was still there, latent, but had been buried alive by a pervasive culture of individualism and consumption. These two themes remain remarkably prescient as we turn to what might be considered the resurrection of social class.

The Great British class survey: a bourdieusian approach to elites

It was against this backdrop that in 2010 the BBC proposed a major new survey seeking to shed light on contemporary understandings of the role and relevance of social class in the UK in the early twenty-first century. The pitch to the public neatly summarised the significance of this:

> It's said that the British are obsessed with class, but does the traditional hierarchy of 'working', 'middle' and 'upper' class really exist anymore? And does social class even matter in 21st century Britain? (BBC, 2016).

The BBC Science division had approached the sociologists Mike Savage and Fiona Devine to devise a web survey aimed at addressing these questions, based upon the view of social class propounded by the French sociologist Pierre Bourdieu. In *Distinction* (1984), Bourdieu argued that social class was more than a matter of economic circumstances or status but rather was the aggregate effect of our differential holdings of various "capitals", of which the economic was just one dimension, to be considered alongside our stocks of "cultural" and "social" capital. Cultural capital is defined as our activities, preferences and interests and Bourdieu argued that rather than being subsidiary to our understanding of how social class works, it should in fact be central to it. He argued that our engagement with particular forms of "legitimate culture" such as going to museums or the theatre not only enhance our social standing, they also have direct benefits in terms of facilitating a successful middle class negotiation of the school system because of the ways in which such established forms of culture are deeply embedded in educational curricula, ways in which more popular forms of cultural engagement simply are not (Savage et al., 2015, p. 95). In *Distinction* Bourdieu's research found that the French elite of the 1960s were set apart from the population not simply by their economic status but by their ability to appreciate at an abstract level, these forms of culture, and by extension to set themselves apart from the forms of culture enjoyed by the masses, as popular and vulgar (Savage et al., 2015, p. 96). For these reasons, the survey dedicated a considerable amount of space to the identification of cultural capital resources; of roughly 250 direct variables available to the research team in the original dataset, 88 of these related to issues of cultural taste and participation. Respondents were asked about their pastimes, tastes in food and music, as well as about the sorts of holidays that they took. In addition to their preferences for each of these various types of activity, they were asked to identify the frequency with which they undertook them, providing considerable data granularity.

From an intellectual standpoint, it is also clear why such a focus was central to an understanding of formation and reproduction of a contemporary elite. A growing evidence base is beginning to emerge around the power of cultural capital in internally-stratifying the elite. Recent research using the *GBCS* has identified a "class ceiling" in operation in the UK's elite occupations in which cultural capital is pivotal (Friedman, Laurison, & Miles, 2015), whilst work by Laurison and Friedman (2016) on the representative Labour Force Survey has found that over 50% of the class pay gap in higher occupations lie unexplained by the available indicators contained therein and that future research needs to focus on the role of the intergenerational transmission of wealth alongside the analysis of social networks and cultural capital. This has been echoed by the Social Mobility and Child Poverty Commission (SMCPC, 2014; 2015a). Their 2014 report found that a small number of prestigious universities dominated access to the most elite roles in British politics, the media and legal sectors, while more recently (Social Mobility and Child Poverty Commission [SMCPC], 2015b) they have identified that a "glass floor" operates to exclude bright working class applicants based on attributes which map precisely onto Bourdieusian notions of social and cultural capital.

Turning then to social capital this also received extensive attention in the survey mechanism for the light it could shed upon the UK's class divides and the upper strata of British society in particular. This was analysed by asking respondents to the survey to identify from a list of 37 different occupational groups ranging from chief executives to call centre workers, those whom they knew socially from amongst their friends, family and acquaintances. Each of these occupations was then given a score based upon the Cambridge Social Interaction and Stratification Scale (CAMSIS), which ranked them according to the level of social interactions between different occupations and the social structures which underpinned these. This approach to the quantitative analysis of social capital is commonly known in the social sciences as a "Lin position generator" following the highly-influential work of Nan Lin (2001) in this field. The resulting data from the *GBCS* position generator could be used and interpreted in different ways to analyse the breadth, size or status of an individual's social networks. For the purposes of analysing the social dynamics behind class structure and the elite in particular, the aggregate status of each respondent's social networks has been the most useful variable derived this area of the dataset. Again, this has solid theoretical underpinnings in terms of sociological theory, drawing upon Mark Grannovetter's (1973) seminal work in a Boston suburb on the "strength of weak ties" between individuals in learning of job opportunities and leveraging economic advantage more generally. In a contemporary and more specifically elite context, such findings still clearly echo in our contemporary concerns around the power of elites to sustain and consolidate their social positions through access to professional networks and specifically, unpaid internships for their offspring. Thus, the SMCPC report,

> Despite increasing recognition that internships should be advertised openly and fairly paid, many are still secured through informal connections and remain unpaid...We are a long

> way from access to internships being based on what, not who, you know (Social Mobility and Child Poverty Commission [SMCPC], 2015b, pp. 105 & 107).

In contrast to the abundance of data available through the survey to address cultural and social capital, the questions addressing the final component, economic capital, were far more constricted. Three questions were asked here: on household income, savings and property value. Clearly, despite the Bourdieusian focus on social and cultural capitals a measure of economic resources forms a key component in our understanding of inequalities and how classes, and elites in particular, are constituted. It is economic inequalities and specifically the inter-generational transfer of wealth and the higher long-run returns on capital which Piketty (2014) identifies has led to the increasing residualisation over time of economic resources within a progressively smaller proportion of the population. Nonetheless, we must also recognise that it is not merely economic resources alone which are heritable and therefore whilst it is important to recognise the significance of the intergenerational transmission of economic advantage, we also need to be clear that a Bourdieusian approach allows for us also to recognise how "soft" assets in the forms of social and cultural capital are also passed between generations, and hence inform the *GBCS* approach to social class. Bourdieu was very clear on this: in his essay on "The Forms of Capital" (1986) he refers repeatedly to the inherited character of social capital, whilst in his work on education (Bourdieu & Passeron, 1979) he again made intergenenerational transmission central to his argument. More generally, the reproduction and inheritance of such soft assets is central to his key concept of *habitus* (1977) in which individuals develop and embody particular worldviews, tendencies and dispositions through everyday practices and interactions. The *GBCS Elite* is therefore a much more theoretically complex formulation.

The primacy of space and specifically, property, is becoming ever more prominent in analyses of elites. In part this could be seen as a sort of superficial manifestation of our popular obsessions with bling and celebrity lives, but at a deeper, more conceptual level, we can view this as a very precise case of "the spatialisation of class". For Burrows (2013) this is both an empirical observation and a methodological exhortation; the accumulation of property by the super-rich in particular parts of London is an exercise in both economic and cultural capital accumulation, it is also a development upon which we need to focus intellectual energy.

For the middle reaches of the class structure, the need to place space much more centrally in our understanding of class identities receives expression in the notion of "elective belonging", elaborated by Savage et al. (2005, p. 29):

> Belonging is not to a fixed community, with the notion of closed boundaries, but it is more fluid, seeing places as sites for performing identities. Individuals attach their own biography to their 'chosen' residential locations so that they tell stories that indicate how their arrival and subsequent settlement is appropriate to their sense of themselves.

Whilst at the other end of the class spectrum the idea of choice is absent from the class-space relationship due to the lack of economic capital that these individuals hold. In this context we have increasingly seen class stigma being expressed within spatial terms of reference and notably in the demonization of social housing estates and those that inhabit them. The increasing iniquity and unsustainability of our housing market have driven this spatialisation of social class, with Jones (2011, p. 35) arguing that;

> over two thirds of those living in social housing belong to the poorest two-fifths of the population, Nearly half of social housing is located in the poorest firth of neighbourhoods. things have certainly changed compared to thirty years ago, when a staggering 20 percent of the richest tenth of the population lived in social housing.

In this way, we therefore need to understand the crucial role that property plays in our understanding of the elite not simply in terms of a cadre perpetually pulling away from the rest of the population, true as this may be. Rather, we need to understand the literal and metaphorical place of the elite as part of a wider process of polarisation in housing, which is acting to radically distend and aggravate class distinctions in the UK and beyond.

The Great British class survey elite: space and capitals

Due in no small part to an extensive publicity campaign, the *GBCS* elicited an enormous and quite remarkable level of response from the general public, with 177,383 responses over the period between the survey's launch in January 2011 and the publication of the research paper in *Sociology* annunciating the initial findings from the data (Savage et al., 2013). Of those responses, nearly 103,000 or 58% occurred in the week immediately following the survey's launch on primetime television and 147,000 or 83% occurred in the month following. The survey remained open until June 2013, during which period the overall sample approximately doubled with a further 160,000 in the 2nd wave, bring the overall sample up to around 320,000.

The 1st wave data were then analysed and a "new model of social class" was proposed based upon variables extracted from the data and designed to operationalise the Bourdieusian capitals discussed above. The new model, derived from latent class analysis, identified a new class structure in the UK in which this *GBCS Elite* constituted approximately seven percent of the population. Both the methodology and the model are discussed in detail in the 2013 research paper (Savage et al., 2013) and indeed have been extensively discussed elsewhere within and beyond academia. The focus in this paper is on the methodological lessons and implications of this extraordinary experiment.

Uniquely, for a survey of this scope, the categorical granularity is matched by considerable spatial detail. The remarkable response rate meant that it was possible to map this new Bourdieusian *Elite* at a high spatial resolution. Figure 1 shows the distribution of the *GBCS Elite* at local authority level for Great Britain and at the level of the county and Belfast Metropolitan Area for Northern Ireland using an "observed versus expected" methodology. This works by highlighting the difference from a baseline of completely even distribution across every spatial unit, relative to the participation level of each unit in the survey itself. Figure 1 shows that the *Elite* are disproportionately concentrated within London and the "home counties" of South East England, with, in general terms, a clear trend towards decline in numbers as distance from the capital increases. At face value these reflect long-established and historic patterns of class settlement in the UK (Dorling, 2004; Howell, 2004) but this geography also clearly resonates with contemporary political debates around the growing spatial and sectoral divides in the UK economy, with a growing concentration of wealth in and around London based primarily on the financial services sector, set against a

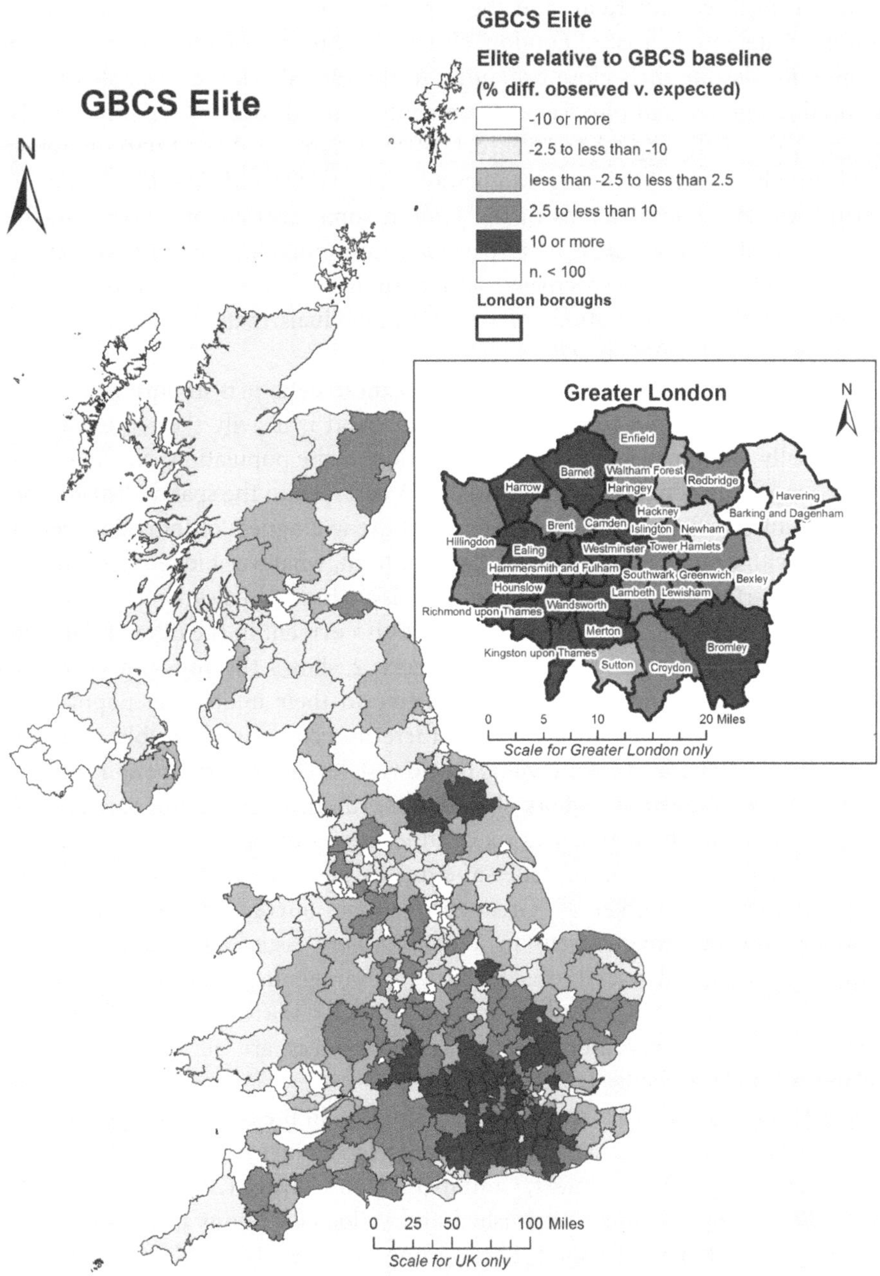

Figure 1. The spatial distribution of the GBCS Elite (Source: GBCS).

progressive decline in the old manufacturing heartlands of the north. Indeed, such is the recognition of the seriousness of these profound imbalances in the UK's social and economic structure that even the Conservative government have acknowledged they must be addressed through their "Northern Powerhouse" initiative (Osbourne, 2014). It is also important to note that the maps do also shed light on some of the regional nuances which reflect the UK's class and economic landscape. For example, the Isle of Thanet in north-eastern Kent, and the Thames Estuary region more generally, which lies immediately to the east of London, clearly show up as areas of under-representation for the *Elite*, despite their close proximity to the capital. The remarkable disparities in the social, economic and physical environments of the districts lying immediately to the east and west of the capital are well described in Iain Sinclair's (2002) psychogeographical circuit of the M25 ring motorway, *London Orbital*. Whilst in the north of England, we see concentrations of the *Elite* in some districts of North Yorkshire and in North Cheshire. Both areas lie in close and commutable proximity to major cities – Leeds and Manchester respectively, and both have come to be known as "golden triangles" for the concentrations of wealthy individuals that lie within the vertices of these (Savage et al., 2015, p. 279).

Yet, despite the detail available from all of these data, it is important to make clear that the *GBCS* was not a representative sample and is heavily skewed towards a more educationally and financially advantaged section of the population. Nor has it captured equally across this population; the main *GBCS* dataset and the spatial distribution of the *Elite* we witness will be, in part, an artefact of the willingness of individual respondents to engage with a survey on social class, and in term, whatever hidden determinants that decision to participate is conditional on. This issue has been clearly acknowledged from the outset (Savage et al., 2013). This skew was also evident in the differential patterns of participation by location. The table in Figure 2 shows the top and bottom twenty districts by participation in the survey relative to their underlying population bases. Nonetheless and skew notwithstanding, there are clear and sensible patterns to be discerned here. The areas witnessing the lowest levels of participation are predominantly in the ex-industrial regions of Wales, Northern England and Scotland's Central Belt, with all of Northern Ireland's six counties appearing also. The last of these can be explained by the interaction of issues of ethnicity and class in Northern Ireland, where the questionnaire's moniker of "Great *British* Class Survey" almost certainly acted to suppress nationalist involvement in a polity where issues of national identity have historically triumphed over those of class (Cunningham & Gregory, 2014; Gregory, Cunningham, Lloyd, Shuttleworth, & Ell, 2013). At the other end of the spectrum, seventeen of the twenty most over-represented areas are all in the South East of England, with eleven being London boroughs.

Despite these imbalances, as Savage (2015) notes, in the case of studying elite groups, the particular nature of the sample skew presents a distinct advantage. Moreover, and in a specifically geographical context, Cunningham and Savage (2015) have found that the *GBCS Elite* closely resembles established and validated notions of a social elite in the form of the National Statistic's Socio-economic Classification's (NS-SeC) Class I, comprised of higher managers and professionals, representing about 12% of the UK population at the last census. This is also notwithstanding the fact that these alternate conceptions of an elite have been based on radically different conceptual and methodological approaches. What the skew clearly shows is that there was a particularly strong appetite for the survey in more affluent areas, and specifically in those parts of the UK where the *Elite* could be found in the greatest numbers. This is further borne out in Figure 3, which is a scatterplot showing the relationship between participation in

Rank	Lowest response	Highest response
1	Tyrone	City of London
2	Fermanagh	Cambridge
3	Armagh	Oxford
4	Blaenau Gwent	Islington
5	Barking and Dagenham	York
6	Bolsover	Hammersmith and Fulham
7	Londonderry	Camden
8	Harlow	Lambeth
9	Stoke-on-Trent	Richmond upon Thames
10	Antrim	Edinburgh, City of
11	Knowsley	Wandsworth
12	Down	South Cambridgeshire
13	Copeland	Southwark
14	Walsall	Bristol, City of
15	Blackpool	St Albans
16	Wolverhampton	Westminster
17	Castle Point	Reading
18	North Lanarkshire	Tower Hamlets
19	Barnsley	Hackney
20	Sandwell	Brighton and Hove

Figure 2. Areas of highest and lowest response to the GBCS across the UK (Source: GBCS).

the survey and the proportion of the sample within each area that were classified as belonging to the *Elite*. The outliers of extreme levels of both participation in the survey and *Elite* composition here are the City of London, Oxford, Cambridge, York and Islington. With these *Elite* geographies we need to consider what is captured by the GBCS mechanism. It is not simply a geography of occupation, affluence or accumulation which we might usually associate with a social elite, but something altogether more complex, nuanced and revealing: a multi-dimensional geography of cultural tastes, social ties allied to economic power. The very strength of this survey lies in precisely the fact that it *is* self-selecting. We are excavating here the geographies of active identification and "elective belonging" (Savage et al., 2005). The sample size means that we can attend to these geographies in remarkable detail. Figure 4 demonstrates this with reference to some of the UK's major cities. Figure 4(a) shows the spatial distribution of the *Elite* in London. The areas that show the greatest concentrations are in places such as Hampstead, Crouch End, Islington, Hackney, extending into the central London neighbourhoods of Bloomsbury and Regent's Park, with pockets in the Docklands and Shoreditch. In South London, areas such as Brixton and Clapham come to the fore. Yet areas of long-held affluence are relatively under-represented in this picture: places such as Chelsea and Mayfair. Turning to Manchester (Figure 4(b)), the areas which emerge are the city centre and key middle-class neighbourhoods to the south of it – Chorlton-Cum-Hardy, Withington and Didsbury.

There is a clear pattern in these geographies which points firmly towards the power of particular high-end urban "villages". In London, these villages take prominence over some a few better-known and perhaps more obvious areas of affluence. They reflect a geography of gentrification, or more accurately, "super-gentrification" in urban space

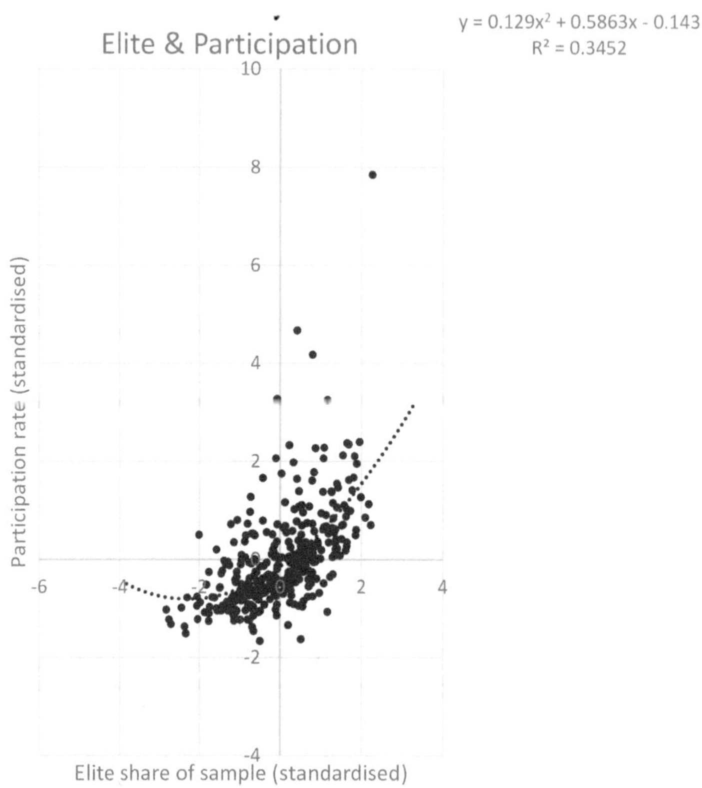

Figure 3. Correlation between Elite composition and participation in the GBCS by spatial unit (Source: GBCS).

(Butler & Lees, 2006). We might then read that as a shortcoming of the utility and accuracy of this *Elite* as a grouping, but this is to neglect the particular way in which it is formulated. The *GBCS Elite* is not simply a financial cadre but represents the aggregate of a complex and complementary stock of social, cultural and economic resources. The sorts of neighbourhoods which the class survey data point to are those where these three forms of capital have most accumulated. They are spaces where very high economic capital are to be found within networks of culturally engaged and highly self-aware, self-confident, high status social networks. In contrast, those areas of London which we might traditionally conceive of as "elite" such as Mayfair and Holland Park, are increasingly characterised, yes, by extreme – indeed stratospheric levels of wealth, but as Atkinson (2015) identifies, also by a lack of social interaction and cultural engagement. The turning of these areas into arenas of absentee super-rich property speculation is leading to new and fascinating forms of social desertification and unrest between the relentless waves of global mineral wealth and the beleaguered "old" money populations (Webber & Burrows, 2015). It is important to note then, that those locales of "new" and "old" money elites do not appear with the levels of intensity which we might otherwise expect. Perhaps these are groups to whom seemingly parochial and arcane debates on social class appear either irrelevant, or following

Figure 4. Geographies of the a) London and b) Manchester Elite (Source: GBCS).

Evelyn Waugh's chastisement of Nancy Mitford over the "u"/'non-u' controversy of the 1950s, still crass and distasteful (Savage et al., 2015).

Yet, if we think in a little more depth about how urban space cleaves so effectively our differing conceptions of elites, it is also possible to use the data to reflect on how our own *GBCS Elite* is also very effectively stratified across the city. Figure 5 demonstrates this powerfully using some key indicators. The analysis has broken Greater London down into 2.5 kilometre wide bands radiating out from the Bank of England located right in the heart of London's financial district. The mean standard scores for three key indicators of each of the three forms of capital has then been generated for all of the *Elite* residents within each bandwidth. This shows some striking patterns. For the *Elite* in London there is a clear and consistent spatial hierarchy across all forms of capital. In terms of all three indicators, there is a clear distance decay effect for members of the *Elite* as proximity to the inner city declines. Higher incomes, higher status social networks higher cultural capital are clearly associated with central London. The analysis underlines the ways in which very different capitals are reinforced and reflected by one other across space, representing an evident geography of "elite distinction" (Daloz, 2009).

So whilst it is clear that there exists pronounced spatial stratification within the *Elite* how does the *Elite* relate to the wider population across urban space? The uniqueness of the *GBCS* means that there are no comparable datasets for the measures of social and cultural capital available from the BBC survey material. However, we can draw on representative data sources for the economic indicators. Figure 6 uses official statistics on annual wages derived Office for National Statistics' (ONS) Annual Survey of Hours

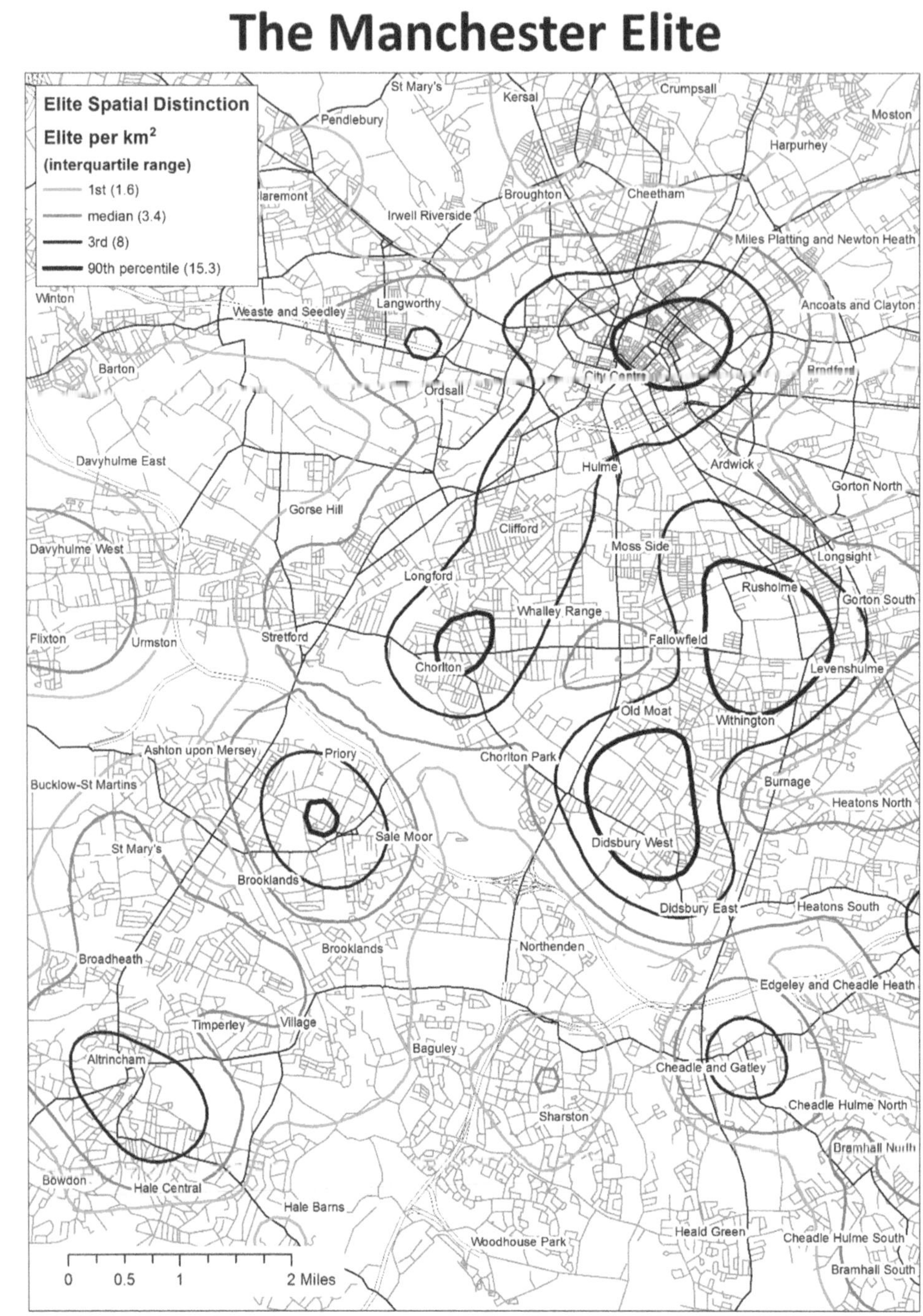

Figure 4. (Continued).

and Income (ASHE) for the ten most populous unitary authorities in Great Britain. London has been divided between the inner and outer boroughs using the ONS' official definition due to the substantial disparities in income between the inner and outer sections, a distinction also evident in the *GBCS* data. Figure 6 presents the ratio of

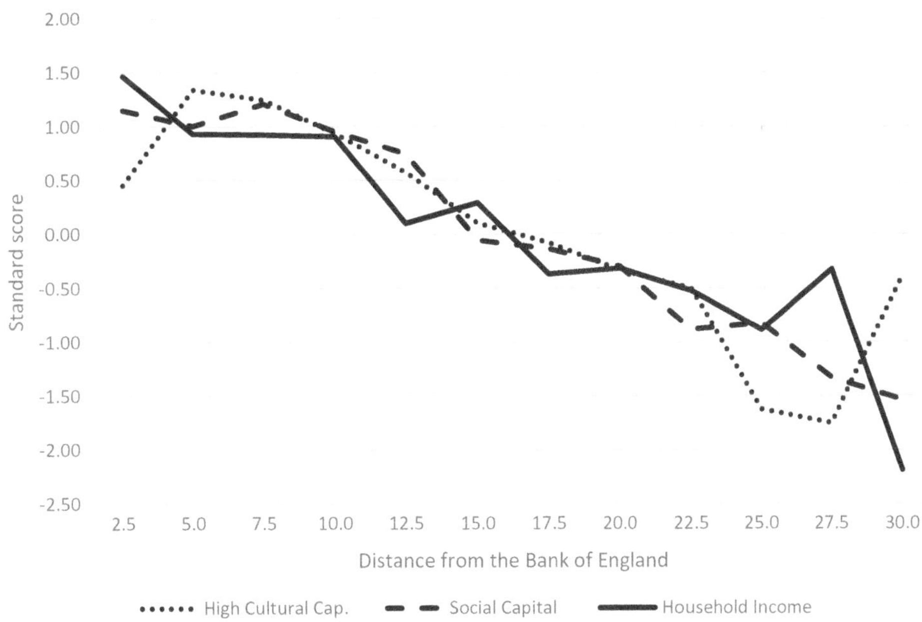

Figure 5. Standard scores for mean household income, mean social capital score and mean high cultural capital score by 2.5 kilometre bandwidth radiating from the Bank of England (Source: GBCS).

annual household income for the *Elite* in the *GBCS* data to individual annual income for those in the 80th percentile of household income from the representative ASHE data. It is thus a means of gauging the economic relationship between the *GBCS Elite* and their wider, representative urban contexts.

Across Great Britain as a whole the ratio of *GBCS Elite* income to ASHE earnings is 2.98:1, but in Manchester the ratio is higher at 3.86:1. The Edinburgh *Elite* displays the lowest level of economic disparity from the wider population at 2.85:1 due to relatively high median wage levels in the city. However, the key point to make about this graph is that there is simply no clear spatial pattern to the unequal relationship between the various urban *Elites* and their particular metropolitan contexts that neatly fits into pre-existing debates on Britain's north-south divide (Martin, 2004) or the economic dominance of London (Cunningham & Savage, 2015). The city *Elite* stand apart but their economic distance from the wider urban populace is nuanced and non-linear.

Responses: the "recursive" elite

As noted, the publication of the "new model" has elicited much interest and critical comment in the intervening years. Those criticisms have ranged across issues of methodology and sample bias (Mills, 2014, 2015), through to issues of whether the model should more accurately be described as one of "stratification" rather than "class" (Bradley, 2015; Skeggs, 2015) due to the weight given to issues of social and cultural capital (Crampton, 2008, pp. 8–9). Or indeed whether such a mechanism is actually helpful at all at a meta-level in terms of understanding of social inequalities or instead does more damage by shifting the focus towards the reification of classed identities

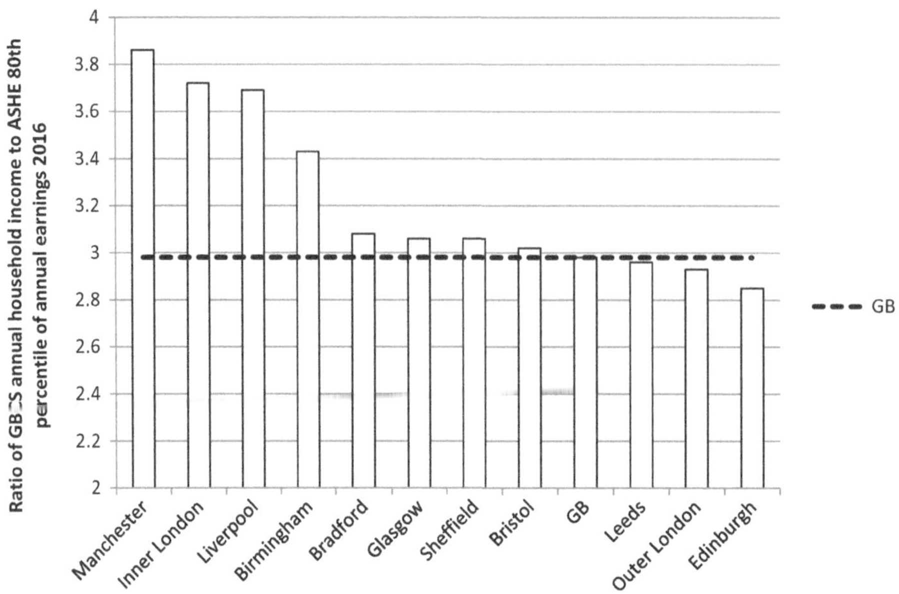

Figure 6. Ratio of GBCS annual household income to ONS Annual Survey of Hours and Earnings (ASHE) 80th percentile of annual income for UK's eleven most populous urban areas (Source: GBCS; ONS ASHE data 2016 (Table 8.7a – Annual Gross Pay – Home Geography).
(Source: GBCS).

rather than the struggle against them (Tyler, 2015) and the enacting of further "symbolic violence" against the weakest in society (Bourdieu, 1984). It was certainly true that in the maelstrom of debate, fuelled by the online character of the survey and the popularity of social media, attention in the immediate aftermath of the "new model" (Savage et al., 2013) focussed heavily on the classifications as little more than a sort of sociological credit score tool. The massively-accessed BBC "class calculator", with seven million unique visitors in the following week, acted as a plaything – a sort of social class "selfie" in which any notions of relationality or implied exploitation between these classes was seemingly lost in a fog of egotistical navel-gazing and self-aggrandising tweets by wealthy, young urbanites in the main, frequently displayed in terms of faux-shock or disapproval (Savage et al., 2015, p. 315).

The sociologist Colin Mills (2014, 2015)) has raised fundamental methodological concerns focusing on the latent class typology which generated the *Elite* as part of the 7-part new model and also on the unrepresentative nature of the *GBCS* dataset which he has argued severely undermines the survey's worth as a tool for social research. It has always been acknowledged that the *GBCS* is heavily skewed in its composition towards those at the upper end of the social spectrum. Savage et al. (2015) make a pragmatic response to these criticisms by showing how it is possible to use these skews effectively to look at those in elite positions for which there are often not enough respondents in national sample surveys to draw any kind of meaningful inferences. It has also been shown how initial work using *GBCS* data has been largely replicated using large scale national sample surveys such as the Labour Force Survey (Laurison & Friedman, 2016),

	GBCS Elite (%)	n
NS-SEC I	58.1	38,153
NS-SEC II	35.0	23,004
Other	6.9	4,566
Total	100.0	42,719

Figure 7. Distribution of GBCS Elite within NS-SeC classes I & II. (Source: GBCS).

thus showing how patterns derived from the *GBCS* can be robust. Furthermore the *GBCS Elite* is not that different *to* notions of an expansive "elite" in the NS-SeC. Figure 7 shows that almost 60% of the *GBCS Elite* fall within NS-SeC I (senior managers & traditional professional occupations), with a further 35% coming under NS-SeC II (modern professional occupations & middle and junior managers). Given the statistical and spatial (Cunningham & Savage, 2015) overlaps between the *GBCS Elite* and nationally-representative and validated measures of class, for pragmatic purposes, the GBCS elite does offer a way of operationalising elites multi-dimensionally, even though we should always recognise the limitation imposed by the self-selecting nature of the *GBCS* dataset.

The overtly geodemographic nature of the latent class model and the archetypes widely circulated and caricatured in April 2013 jarred with traditional academic approaches to class analysis and gave rise to accusations that the *GBCS* experiment was purely an exercise in "data dredging" (Mills, 2014). However, the variables used to construct the MCA were carefully and conceptually specified to measure cultural and social capital. In the case of the *GBCS*, if any theoretical criticism is to made it might be that the experiment was perhaps too hidebound to an *a priori* conceptual framing rather than too little in terms of a Bourdieusian approach to class structure over established conceptual approaches. Notwithstanding the clear limitations of a class structure based only on occupation and employment relations, to have dispensed entirely with such measures in favour of a solely Bourdieusian economic, social and cultural capitals approach might be regarded by some as an unnecessarily radical departure given the primacy and value of occupational based approaches over such a long period. However had some form of composite measure been put forward which included both occupational and Bourdieusian approaches it is likely that this too would have led to accusations of conceptual incoherence.

Another of our critics, the geographer Danny Dorling, whose work on inequality has focussed on a more exclusive definition of an "elite" in his "1%" (2014), made the point that while our classification of an elite was too broad, the concept more generally was problematic as it is inherently "recursive" (Dorling, 2013). As a self-evident truth, there is no real response to the notion of recursivity, so the main contention then lies in how one conceives of an "elite". The explicit conception with the *GBCS Elite* is of a new "ordinary" cadre who share economic, social and cultural privileges beyond the reach of the vast majority of the population (Savage et al., 2015, pp. 303–329). Part of the rationale behind this is in the clear knowledge that wealth has increasingly become concentrated in the hands of an ever smaller proportion of the population; it has been a long term trend, even if the evidence is now that it appears to be accelerating after some decades of relative egalitarianism in the post-War period (Piketty, 2014). Further, despite the fact that such extremes of wealth have long been apparent, it does not appear to have acted as a particularly powerful call for remedial action. The purpose of this experiment was to understand processes of inequality and the mechanisms by which people end up in their given class location and therefore to move beyond the empirical descriptions of economic inequalities alone which have dominated debate up to this point, valid and necessary as those are. In this intellectual context then, the idea of a wider conception of an "elite" in which we can see the ways in which inequalities are replicated and reinforced beyond the economic sphere, appears like a radical departure and one which would seem to resonate with new research which draws attention to the interactive effect of Bourdieusian capitals and the economic power of cultural and social capital resources (SMCPC, 2014, 2015a, 2015b; Friedman et al., 2015).

However, the criticism levelled by Dorling (2013) and Tyler (2015) that in creating this new *Elite* the focus became more about the taxonomy rather than the underlying structures and relationships, is a significant one. This is underlined powerfully in spatial and temporal analysis of the *GBCS* response patterns. Figure 8 shows the composition of the two waves of response in terms of the 2013 "new model" classifications (Savage et al., 2013) and is both a telling and depressing comment on the limits of mass and new medias to speak to the population as a whole. As we know, the data were already heavily skewed towards the most privileged in society, with the *Elite* comprising 21.4% in wave 1 (as opposed to seven percent in the nationally-representative sample). However, in wave 2, the *Elite* proportion leapt even further, now comprising 27% of all respondents in the second collection period. That statistic provides powerful evidence of two mutually reinforcing dynamics: the highly-competitive desire for affirmation on the part of society's "winners" at one end (Daloz, Jean-Pascal, 2009) and the "conditions of governmentality" (Hannah, 2001; Legg, 2005) which act to shape how society's "losers" interact with the mechanisms which maintain their place at the bottom. This becomes even more apparent when we look in detail at the relative participation rates for the *Elite* on the results launch day of 3 April 2013. Figure 9 shows the observed rate of participation for the *GBCS Elite* against the expected rate, were the *Elite* to respond at the same rate as the *GBCS* sample as a whole. Here we see that the *Elite* was heavily over-represented earlier in the day, in the period up to 1pm in

New Class	Wave 1 (%)	Wave 2 (%)
Elite	21.4	27.0
EMC	42.9	39.8
NAW	6.1	5.4
TMC	9.1	8.9
TWC	1.7	1.4
ESW	18.1	16.7
PRE	.8	.8

Figure 8. The class composition of the GBCS in Wave 1 (January 2011 – February 2011) and Wave 2 (March 2011 – June 2013).

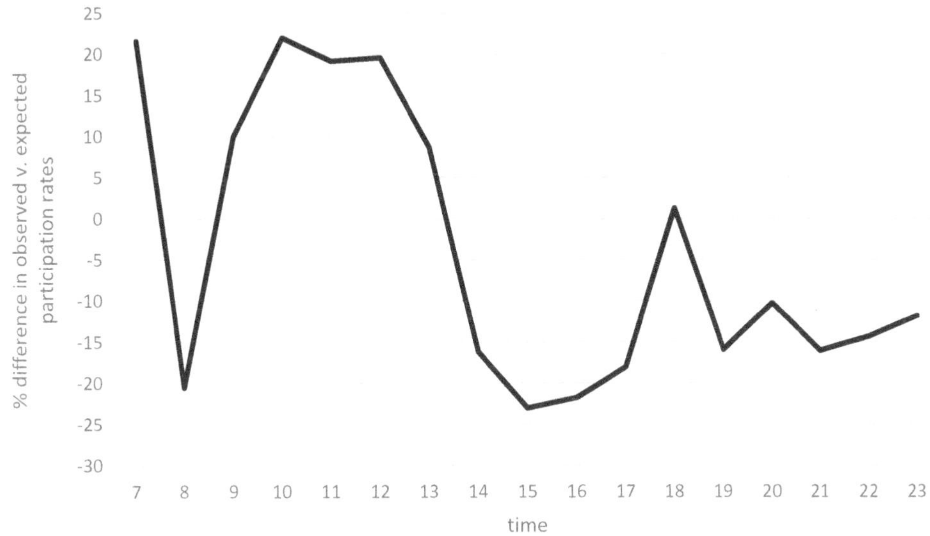

Figure 9. Observed versus expected participation in the GBCS by the Elite by hour on 3 April 2013 (Source: GBCS).

the afternoon, with the only exception being the hour between 8am and 9am when most would have been commuting to places of work or on school drop offs, etc. It seems that this *Elite* were clearly anxious to acquire and affirm their status early on.

And again these patterns are reflected when we look to the spatial distribution of patterns of response. As we already noted from Figure 3, the more elite an area was, the more likely its denizens were to participate in the survey. Figure 10(a,b) maps the response rate for the two separate waves. Figure 10(a) shows that the areas of disproportionately higher response were also the most affluent parts of the country: west and south-west London and the Home Counties surrounding the capital. Far lower levels of response that expected were registered in Northern Ireland, Scotland, Wales and many parts of northern England. We can see from Figure 10(b) that this pattern was replicated in wave 2, where there no great corrective from previously under-represented parts of the country. The lack of change between the two waves underlines the skewed spatial enthusiasm and apathy with which the *GBCS* was greeted, and is stark testament to the distinctly classed and spatialized nature of the metropolitan discussion and critique which accompanied the genesis of the *GBCS Elite* and its other class categories. It also appears to speak directly to the work of Jean-Pascal Daloz (2009, p. 28) who has argued that intra-group competition has always been a central, yet neglected aspect of the way in which elites reify their status. Such competition might traditionally have been manifest more in material displays of wealth and ostentation or in social and cultural affectations (Daloz, Jean-Pascal, 2009, p. 69, pp. 103–110), but in the digital world it appears that the online survey mechanism has provided a powerful means of the kind of conscious self-affirmation which he identifies.

Figure 10. Spatial distribution of participation in the GBCS by unitary authority* in: A) Wave 1 (January 2011 – February 2011) and b) Wave 2 (March 2011 – June 2013.
(Source: *GBCS*).
• *Belfast Metropolitan Urban Area and counties for Northern Ireland.*

Conclusion

This contribution has shed light on one specific and contentious incarnation of an elite. The purpose of this paper is to highlight the need to understand the elite as a specific class formation, which comprises distinct resources of social and cultural, as well as economic capital, which set them apart from the population at large. The *GBCS* was a unique and powerful tool for identifying the new *Elite* and the complex relationships between these capital resources, and unlike previous historical conceptions of a British elite, this is a distinctly urban grouping, for whom London acts as a fulcrum. However, there exist pronounced intra-stratal distinctions within the London *Elite*, distinctions which are reinforced across all forms of capital and which are thrown into sharp relief through geographic analysis. These data point to the existence of nested urban villages, operating almost as class cultural ecosystems, in which groups of people with similar levels of economic capital congregate. However, what makes this *Elite* distinctive is the light it sheds beyond those material resources, turning our attention to these privileged spaces of repositories of social and cultural capital as well. Nothwithstanding the limitations of the data which have been discussed here and elsewhere in some detail, it is in this regard that the *GBCS* reveals its research potential. Turning beyond London it is evident that simplistic north-south binaries break down when we consider both this *Elite's* spatial distribution and their economic relationship to their wider urban contexts. Yes, the *Elite* are a profoundly southern formation, but they also exist in significant numbers in areas of long established wealth in the north of England and beyond. Furthermore, the economic disparities between the provincial *Elite* in Manchester and the surrounding host district are more pronounced than those in Inner London. More generally there appears to be no clear spatial pattern to the extent of those disparities across the country as a whole.

In addition to its substantive empirical data in terms of the three Bourdieusian capitals, the *GBCS* is also highly revealing at a methodological and ontological level as displayed in the pronounced spatial, temporal and socio-economic patterns of engagement which the tool engendered. These patterns of engagement have clearly pointed to the power of the *GBCS* as a tool for the metropolitan *Elite* to affirm their sense of self and to allay their insecurities, whilst for those at the other end of the class spectrum, it has been largely ignored. So whilst the spatial trend in London and elsewhere may be for elites to increasingly shield (Atkinson, 2016) and segregate themselves though the construction of "poor doors" (Osborne, 2014) and subterranean worlds (Webber & Burrows, 2015, pp. 11–12), such secrecy does not extend to the digital sphere, at least not for our "new 'ordinary' *Elite*" (Savage et al., 2015).

Acknowledgements

I am extremely grateful to the Editor and to three anonymous referees for their time and insightful thoughts on this article. I also wish to thank Mike Savage for reading and commenting on an earlier version.

Disclosure statement

No potential conflict of interest was reported by the author.

Funding

This work was supported by the Economic and Social Research Council [RES-577-28-0001].

References

Anderson, Benedict. (1991). *Imagined communities: Reflections on the origins and spread of nationalism* (2nd ed.). London: Verso.
Atkinson, Rowland. (2016). Limited exposure: Social concealment, mobility and engagement with public space by the super-rich in London. *Environment and Planning A, 48(7), 1302-1317.*
Atkinson, Rowland, Parker, Simon, & Burrows, Roger. (2017). Elite formation, power and space in contemporary London. *Theory, Culture and Society (Early Online)*, *34*(5–6), 179–200.
Atkinson, Will. (2010). *Class, individualization and late modernity: In search of the reflexive worker*. Basingstoke: Palgrave.
BBC. (2016). BBC lab UK: The Great British class survey. Retrieved April 2016, 28, from http://www.bbc.co.uk/labuk/experiments/the-great-british-class-survey
BBC News. (1999, 28 September). UK politics: Tony Blair's speech in full. *BBC News* website. Retrieved April 2016, 13, from http://news.bbc.co.uk/1/hi/uk_politics/460009.stm
Beaverstock, Jonathan V., Hubbard, Phil, & Scott, John R. (2004). Getting away with it? Exposing the geographies of the super-rich. *Geoforum*, *35*(4), 401–407.
Beck, Ulrich, & Beck-Gernsheim, Elisabeth. (2002). *Individualisation: Institutionalised individualism and its social and political consequences*. London: Sage.
Bourdieu, Pierre. (1977). *Outline of a theory of practice*. Cambridge: Cambridge University Press.
Bourdieu, Pierre. (1984). *Distinction: A social critique of the judgement of taste*. Cambridge, MA: Harvard University Press.
Bourdieu, Pierre. (1986). The forms of capital. In J. Richardson (Eds.), *Handbook of theory and research for the sociology of education* (pp. 241–258). New York, NY: Greenwood.
Bourdieu, Pierre, & Passeron, Jean-Claude. (1979). *The inheritors: French students and their relation to culture*. Chicago: University of Chicago Press.
Bradley, Harriet. (2015). Class descriptors or class relations? Thoughts towards a critique of Savage et al. *Sociology*, *48*(3), 429–436.
Brown, Laurence, & Cunningham, Niall. (2016). The inner geographies of a migrant gateway: Mapping the built environment and the dynamics of Caribbean mobility in Manchester, 1951–2011. *Social Science History*, *40*(1), 93–120.
Burrows, Roger. (1999). Residential mobility and residualisation in social housing in england. *Journal Of Social Policy*, *28*(1), 27-52. doi:10.1017/S0047279499005474
Burrows, Roger. (2013, December 3). The new gilded ghettoes: The geodemographics of the super-rich. *Discover Society*. Retrieved April 2017, 28, from http://www.discoversociety.org/2013/12/03/the-new-gildedghettos-the-geodemographics-of-the-super-rich/
Butler, Tom, & Lees, Loretta. (2006). Super-gentrification in Barnsbury, London: Globalisation and gentrifying global elites at the neighbourhood level. *Transactions of the Institute of British Geographers*, *31*(4), 467–487.
Butler, Tom with Robson, Garry. (2003). *London calling: The middle classes and the re-making of inner London*. Oxford: Berg.
Crampton, Rosemary. (2008). *Class and Stratification* (3rd ed.). Cambridge: Polity Press.
Cunningham, Niall, & Gregory, Ian. (2014). Hard to miss, easy to blame? Peacelines, interfaces and political deaths during the Northern Ireland Troubles. *Political Geography*, *40*, 64–78.
Daloz, Jean-Pascal. (2009). *The sociology of elite distinction: From theoretical to comparative perspectives*. London: Palgrave Macmillan.
Dodgshon, Robert. (2008). In what way is the world really flat? Debates over geographies of the moment. *Environment and Planning D: Society and Space*, *26*(2), 300–314.

Dorling, Danny. (2013, April 11). What class are you?' *Statistics views* website. Retrieved May 2016, 26, fromhttp://www.statisticsviews.com/details/feature/4582421/What-Class-Are-You.html
Dorling, Danny. (2014). *Inequality and the 1%*. London: Verso.
Dorling, Danny. (2004). Distressed times and areas: Poverty, polarisation and politics in England, 1918–1971. In A.R.H. Baker & M. Billinge (Eds.), *Geographies of England: The North-South divide, material and imagined* (pp. 44–63). Cambridge: Cambridge University Press.
Forrest, Ray, & Murie, Alan. (1983). Residualization and council housing: Aspects of the changing social relations of housing tenure. *Journal of Social Policy*, *12*(4), 453–468.
Friedman, Sam, Laurison, Daniel, & Miles, Andrew. (2015). Breaking the 'Class' ceiling? Social mobility into Britain's Elite occupations. *Sociological Review*, *63*(2), 259–290.
Granovetter, Mark S. (1973). The strength of weak ties. *American Journal of Sociology*, *78*(6), 1360–1380.
Gregory, Ian N., Cunningham, Niall, Lloyd, Christopher D, Shuttleworth, Ian G., & Ell, Paul S. (2013). *Troubled geographies: A spatial history of religion and society in Ireland*. Bloomington, IN: Indiana University Press.
Hannah, Matthew G. (2001). Sampling and the politics of representation in US Census 2000. *Environment and Planning D*, *19*(5), 515–534.
Howell, Philip. (2004). Industry and identity: The North-South divide and the geography of belonging, 1830–1918. In A.R.H. Baker & M. Billinge (Eds.), *Geographies of England: The North-South divide, material and imagined* (pp. 64–87). Cambridge: Cambridge University Press.
Jones, Owen. (2011). *Chavs: The demonisation of the working class*. London: Verso.
Laurison, Daniel, & Friedman, Sam. (2016). The class pay gap in higher professional and managerial occupations. *American Sociological Review*, *81*(4), 668–695.
Legg, Stephen. (2005). Foucault's population geographies: Classifications, biopolitics and governmental spaces. *Population, Space and Place*, *11*(3), 137–156.
Lin, Nan. (2001). *Social capital: A theory of social structure and action*. Cambridge: Cambridge University Press.
Martin, Ronald L. (2004). The contemporary debate over the North-South divide: Images and realities of inequality in late-twentieth-century Britain. In A.R.H. Baker & M. Billinge (Eds.), *Geographies of England: The North-South divide, material and imagined* (pp. 15–43). Cambridge: Cambridge University Press.
Mills, Colin. (2014). The great british class fiasco: a comment on savage et al. *Sociology*, *48*(3), 437–444. doi:10.1177/0038038513519880
Mills, Colin. (2015). The great British class Survey. *Requiescat in Pace, Sociological Review*, *63*(2), 393–399.
Cunningham, Niall & Savage, Mike. (2015). The Secret Garden? Elite Metropolitan Geographies in the Contemporary UK. *Sociological Review*, *63*(2), 321–348.
Osborne, George. (2014a, June 24). We need a Northern powerhouse. Speech delivered by the Chancellor of the Exchequer, Museum of Science and Industry, Manchester. Retrieved May 2016, 3, from https://www.gov.uk/government/speeches/chancellor-we-need-a-northern-powerhouse
Osborne, Hilary. (2014b, 25 July). Poor doors: The segregation of London's inner-city flat dwellers. *Guardian*. Retrieved May 2016, 26, from http://www.theguardian.com/society/2014/jul/25/poor-doors-segregation-london-flats
Pahl, Ray E. (1989). Is the emperor naked? Some questions on the adequacy of sociological theory in urban and regional research. *International Journal of Urban and Regional Research*, *13*(4), 709–720.
Pakulski, Jan. (1993). The dying of class or of marxist class theory? *International Sociology*, *8*(3), 279–292.
Pakulski, Jan, & Walters, Malcolm. (1996). *The death of class*. London: Sage.

Parker, Simon, Uprichard, Emma, & Burrows, Roger. (2007). Class places and place classes: Geodemographics and the spatialization of class. *Information, Communication & Society, 10* (6), 902–921.

Piketty, Thomas. (2014). *Capital in the Twenty-First century*. Cambridge, MA: Harvard University Press.

Power, Anne, & Tunstall, Rebecca. (1995). *Swimming against the tide: Polarisation or progress on 20 unpopular housing estates, 1980–1995*. York: Joseph Rowntree Foundation.

Ray, Larry, & Sayer, Andrew. (1999). Introduction. In L. Ray & A. Sayer (editors), *Culture and economy after the cultural turn* (pp. 1–24). London: Sage.

Rex, John, & Moore, Robert. (1967). *Race, community and conflict: A study of sparkbrook*. London: Institute of Race Relations.

Savage, Mike. (2015). Introduction to elites: From the 'problematic of the proletariat' to a class analysis of 'wealth elites'. *Sociological Review*, 63(2), 223–239.

Savage, Mike, Bagnall, Gaynor, & Longhurst, Brian. (2005). *Globalisation and belonging*. London: SAGE.

Savage, Mike, & Burrows, Roger. (2007). The coming crisis of empirical sociology. *Sociology, 41* (5), 885–899.

Savage, Mike, Cunningham, Niall, Devine, Fiona, Friedman, Sam, Laurison, Daniel, Mckenzie, Lisa, … Wakeling, Paul. (2015). *Social class in the Twenty-First century*. London: Penguin.

Savage, Mike, Devine, Fiona, Cunningham, Niall, Friedman, Sam, Laurison, Daniel, Miles, Andrew, Snee, Helene, & Taylor, Mark. (2015a). On social class, anno 2014. *Sociology, 49* (6), 1011-1030. doi:10.1177/0038038514536635

Savage, Mike, Devine, Fiona, Cunningham, Niall, Taylor, Mark, Yaojun, Li, Hjellbrekke, Johs, … Miles, Andrew. (2013). A new model of social class? Findings from the BBC's great British class survey experiment. *Sociology*, *47*(2), 219–250.

Savage, Mike. (2010). *Identities and Social Change in Britain Since 1941: The Politics of Method*. Oxford: Oxford University Press.

Sinclair, Iain. (2002). *London orbital: A walk around the M25*. London: Granta.

Skeggs, Bev. (2015). Introduction: Stratification or exploitation, domination, dispossession and devaluation? *Sociological Review*, *63*(2), 205–222.

Slater, Tom. (2009). Missing Marcuse: On gentrification and displacement. *City*, *13*(2), 292–311.

Social Mobility and Child Poverty Commission (SMCPC). 2014. Elitist Britain? London: HMSO.

Social Mobility and Child Poverty Commission. (2014). *Elitist Britain?* London: HMSO.

Social Mobility and Child Poverty Commission. (2015a). *A qualitative evaluation of non-educational barriers to the elite professions*. London: HMSO.

Social Mobility and Child Poverty Commission. (2015b). *State of the Nation 2015: Social mobility and child poverty in great Britain*. London: HMSO.

Tanndal, Julia, & Waldenström, Daniel. (2016, April 13). Big Bang financial deregulation and income inequality: Evidence from UK and Japan. *Centre for Economic Policy Research (CEPR) Vox* website. Received April 2016 13, from http://www.voxeu.org/article/financial-deregulation-and-income-inequality

Turner, Alwyn W. (2013). *A classless society: Britain in the 1990s*. London: Aurum Press.

Tyler, Imogen. (2015). Classificatory struggles: Class, culture and inequality in neoliberal times. *Sociological Review*, *63*(2), 493–511.

Webber, Richard. (2007). The metropolitan habitus: Its manifestations, locations and consumer profiles. *Environment and Planning A*, *39*(1), 182–207.

Webber, Richard, & Burrows, Roger. (2015). Life in an alpha territory: Discontinuity and conflict in an elite London 'village'. *Urban Studies*, 53(15), 3139-3154.

Weeden, Kim, & Grusky, David. (2005). The case for a new class map. *American Journal of Sociology*, *111*(1), 141–212.

Weeden, Kim, & Grusky, David. (2012). The three worlds of inequality. *American Journal of Sociology*, *117*(6), 1723–1785.

Wilkinson, Richard, & Pickett, Kate. (2010). *The spirit level: Why more equal societies almost always do better*. London: Allen Lane.

The challenge of researching "partial exit" and "rootedness" among upper-middle classes in European cities

Alberta Andreotti, Patrick Le Galès and Francisco Javier Moreno-Fuentes

ABSTRACT

The rise of mobilities has paved the way for important changes within cities and the possibility for urban upper-middle classes to exit from their cities and national societies, disinvesting on them, while still taking profits in a process of deterritorialization with important consequences at the collective level. Investigating the extent to which upper-middle classes shaped and are still rooted (or not) in their cities is a challenge, even more when international comparison is involved. In this paper we focus on two methodological aspects: 1) the different social meaning that upper-middle classes and the social groups composing them can have considering three Western European countries and four cities: Paris, Lyon, Milan and Madrid. 2) How we generated our data through personal interviews with European managers, and the conceptual framework that informed the understanding of rootedness, exit or partial exit of our managers.

After a period of relative lack of interest on the study of elites, academic research on these groups is flourishing again, spurred by increasing economic inequalities spreading across all Western capitalist societies (but also in many other parts of the world), as discussed by a number of distinguished scholars (Atkinson, 2015; Davis & Williams, 2017; Piketty, 2014; Savage & Williams, 2008).

Investigating the global elite to go beyond the quasi-mythological representation of a homogeneous, hyper-mobile, selfish, "deterritorialized" group, strategically deciding the fate of the world while seating around a table, can certainly be a very fruitful research endeavor (Atkinson, Parker, & Burrows, 2017; Burrows, Webber, & Atkinson, 2017). But the one percent of the super-rich does not share much with the rest of their co-nationals, so studying them does not help us understanding the way in which the majority of the population is living and adjusting their lives, practices, beliefs and values to the changes brought about by globalization in contemporary Western societies. A better understanding of these processes may actually come from the analysis of the way in which larger social groups (although still at the upper ladder of the social stratification system) deal with those transformations, and with the challenges they create.

Our research focuses on the upper-middle class (a group located around the 5–15% of the income hierarchy) analyzing their role in the transformations currently experienced by contemporary European cities. The main purpose of this research is to explore the extent to which the upper-middle classes constitute the backbone of the "transnational mobile society". This paper also looks at the extent to which they may be in the process of becoming "deterritorialized", abandoning public space, and avoiding interactions with other social groups within their cities as a response to the challenges brought about by globalization. According to what we already know about their income, education and mobility, they may constitute a new European urban bourgeoisie.

The process of globalization refers to the progressive growth of interactions and exchanges in the world, or at least a significant part of it in which mobility (of people, capital, information, goods, and images) constitutes a major driver for socio-economical change. Individuals and objects are (or at least are supposed to be) increasingly on the move (Kaufman, 2003). Thanks to new technologies and the reduction of travel costs, more people can indeed afford to travel abroad for tourism, studies or professional purposes than ever before (Eurostat, 2015; Vandenbrande, 2006). The process of European integration significantly reduces the costs of mobility of European citizens within the Union, facilitating the development of strategies of internationalization through the granting of a series of key rights, and by greatly simplifying the administrative procedures to be able to work, study or live abroad (Díez-Medrano, 2010; Recchi, 2015; Recchi & Favell, 2009). This dynamic raises serious methodological questions.

Besides mobility in the real world, increasing "virtual mobility" has also come to have very specific effects on individual's lives and behaviors (Andreotti, Le Galès, & Moreno-Fuentes, 2014; Elliott & Urry, 2010). The hyper-connectivity associated to the increasing use of information technologies has made possible to instantly and virtually cross national borders to work, consume, enjoy, settle down, invest or hide money, as well as to maintain social interactions (Jansson, 2016). Groups and individuals enjoy greater opportunities to be mobile at different levels, potentially contributing to the gradual emergence of a partialy transnational/global/cosmopolitan society (Robinson & Harris, 2000; Sklair, 2001, 2005). This "transnational mobile society" is supposed to bring about a process of deterritorialization affecting individuals and organizations, straining and weakening both national and local societies in favor of a more global sense of belonging (Urry, 2000). Bypassing national and local constraints in order to obtain greater (collective or individual) benefits, these mobile groups may contribute to blur national hierarchies, thus threatening national and local social cohesion. Therefore, social groups and individuals within the nation-state are less interdependent on each other than at any other point in the modern past (Bartolini, 2005). "Deterritorialized" individuals may develop weaker attachments and commitments to place and to others within the community where they live, and this may have crucial implications at the collective level (for instance for social solidarity and wealth redistribution).

However, nobody is mobile all the time and to the same extent. Arguments elaborating on the emergence of a "transnational mobile society" point towards the upper-middle classes as the key social actor. A variety of empirical studies from different perspectives show that some social and professional groups are indeed more likely to be mobile than others. The literature on business travel and reorganization of labor highlights the rise in the volume and frequency of international business travels for some groups, notably managers, senior consultants, and finance lawyers of the service sector (Beaverstock et al., 2011). Surveys on mobility still point at the existence of a significant differences between social classes in the likelihood of travelling, with a clear advantage for the upper classes (Díez-Medrano, 2010; Gusftason, 2009; Mau, 2010; Vandenbrande, 2006).

Upper-middle classes are highly educated, and tend to occupy better positions in the labour market (Davis & Williams, 2017), so they are considerably more likely to have access to the linguistic, cultural, social and cognitive skills and competences necessary to be internationally mobile (Recchi & Favell, 2009). Furthermore, they are able to take full advantage of mobility opportunities thanks to those competences (Fligstein, 2008; Larsen, Urry, & Axhausen, 2006).

To investigate the transnationalization and deterritorialization of upper-middle classes we focused on the ones who stayed in, or returned to their home country.[1] Based on the empirical evidence generated through 480 interviews with managers living in Paris, Lyon, Madrid and Milan, and combining different methods and analytical techniques, our results stress that upper-middle classes living in these four Western European cities are rather transnational (although their practices differ quite significantly in this respect[2]), and they are actually not becoming "deterritorialized" or uprooted. Rather, their strategies within the city show that they skillfully act on different scale levels, putting into practice strategies of "partial exit", choosing where and to what extent they get involved in local matters, while carefully playing the game of distance and proximity with other social groups. This way, they reproduce and reaffirm their relatively privileged position in their national societies, while gradually emerging as a European upper-middle class in the making.

In this article we do not delve into the actual empirical findings of that research (Andreotti et al., 2014). This paper focuses on the conceptual and methodological questions that had to be addressed when designing that research. It provides some useful insights into the study of urban elites. Three specific dimensions are developed in this paper: 1) how the boundaries of the social groups constituting the upper-middle classes in continental European cities were defined in our research; 2) the way in which we approached the combination of transnational practices and "rootedness" in their local environments put in place by the upper-middle classes; 3) finally, elaborating on how we conceptualized and analyzed the "exit" and "partial exit" strategies adopted by these groups at the urban level.

These conceptual and methodological aspects were informed and guided by these theoretical considerations.

(1) Continental European societies have quite distinctive features: a deep territorialization of social life, family bonds, politics, and economic relations (Crouch, 1999; Mendras, 1997); a long history of medium size cities, which remains important up until today (Bagnasco and Le Galès, 2000; Kaelble, 1990); as well as the role of longstanding urbanized *bourgeoisies*, and their heavy influence in these cities. The rise of urban bourgeoisies, and the social groups composing them, was a smooth and variable process in the different socio-political and territorial contexts of continental European societies. Understanding these features requires taking into consideration the specificities of each society, without falling into the potential trap of "methodological nationalism".

(2) Mobility should not be studied independently from other social features and factors. The impact of mobility in European societies is better understood in relation to the "rootedness" of individuals within families, social groups, cities, welfare regimes and national societies. Mobility and mobile lives only make sense in relation to local lives and neighborhoods. The combination of these two dimensions prevents us from perceiving a distorted image of social change where there actually is not much of it. Following a well-established tradition in urban studies (Burawoy, 2005; Hannerz, 2003; Kennedy, 2007; Savage, Bagnall, & Longhurst, 2005), we combined the analysis of transnational mobility practices, with considerably more stable residential realities.

(3) Class dynamics in contemporary societies are not just reflected in work relations, or in sharing common values and attitudes (traditional indicators of class), but also in residential choices and practices, as well as in the nature and intensity of interaction in social networks. The residential dimension, together with the intensity of involvement within networks of social interaction has become central in the process of definition of the upper-middle classes' social position, and their social reproduction, allowing these groups special access to other important fields (work, culture and crucially, education) in which they operate (Atkinson & Gary, 2005; Favell & Recchi, 2011; Savage et al., 2005).

This article starts by identifying the different groups of upper-middle classes in European cities, explaining how this category was used in our research, and how informants were recruited to participate in it. In the second section particular attention is paid to the way in which urban upper-middle class managers develop residential strategies, and how those choices contribute to shape the city. The third section focuses on the way in which the concept of "partial exit" was conceptualized and operationalized in our research, analyzing the dynamics of "deterritorialization" of those groups, and particularly reflecting on the balances of global and local practices developed by these upper-middle class managers. The final section provides some concluding remarks on the relevance of studying urban upper-middles classes for understanding the evolution of contemporary cities.

1. Searching for the upper-middle classes in european cities

Academic debates on the emergence of a transnational mobile society raise some major theoretical and methodological questions. Who are these social groups particularly active in the process of increasing transnationalization of their societies? Which are their main features? These questions point to the definition of the urban upper-middle classes that constitute the focus of our research, to the operationalization of the category of upper-middle classes in a comparative setting, and to how and through what channels we selected the urban upper-middle class individuals included in our sample.

Comparing socio-economic classes in different countries constitutes a very complex challenge since representation, dynamics and measurements are deeply "rooted" in national trajectories, social systems and social contracts, even when these are slowly being eroded (Zunz, Schoppa, & Hiwatari, 2002).

In the European context, States and societies have been in the making over several centuries, and different socio-historical processes have shaped the rise of the middle classes in each society: the industrial revolution, the consolidation of the State, its bureaucracy and the welfare states, the rise of industrial capitalism, and more recently the processes of tertiarization and de-industrialization of the economy (Crouch, 1999; Esping-Andersen, 1990; Therborn, 1995). As a result of these complex processes, four social groups can be considered to constitute the main components of the upper-middle classes in European cities: entrepreneurs accumulating capitals with the industrial revolution; bureaucrats and high-ranking State employees accumulating power resources with the consolidation of the nation state; managers emerged with the consolidation of vertically integrated large industrial firms; and old and new professionals (doctors, lawyers, accountants, as well as consultants, finance brokers, media managers, executives, etc.) gaining autonomy and accumulating economic and other socially relevant resources.

Acknowledging the complexity and heterogeneity of these social groups, our research focused on one specific segment of the upper-middle classes: managers, from both the private and public sectors. Managers constitute a particularly interesting social group that has been relatively neglected in the literature and in empirical studies focusing on the upper-middle classes, representing a peculiar relation to the ownership of the means of production. Gouldner (1979) identified managers as a new social class, a technical intelligentsia in capitalist industrial society, constantly in tension with political elites on the one hand, and with those inheriting capital on the other. This author's key insight points to the conflicts between managers, who have the knowledge and expertise to run firms and may challenge the power and authority of the owners of capital, and this latest group who constitutes the old dominant class, and whose power is based upon inherited privileges and wealth. Over the last few decades, managers have experienced a rapid growth, acquiring economic and social assets so as to gain from changing scales, as well as from the processes of Europeanization and globalization (Favell & Guiraudon, 2011; Mau, 2010). For these reasons, they are the prime suspect to have an interest in blurring national hierarchies (i.e. increasing the relative importance attributed to international experiences in meritocratic selection processes; supporting the increasing role of supranational organizations in the governance of an increasing number of social, economic and political spheres; embracing the development of international cultural patterns, etc.) even though they themselves have become profoundly "rooted" in their national societies, "caged" by war and welfare as Mann put it (Mann, 2013).

1.1. *Operationalizing managers: national social stratification patterns and statistical categories*

Despite the fact that the category of managers was mainly developed after WWII in the countries included in this study, important differences in timing and conceptualization of this category clearly emerged among them. France is the country where the category of manager developed earlier, where it constitutes the highest proportion of the total working population, and the social prestige associated to this category is higher. Since the 1960s, "*les cadres*", with their social status, privileges and rights, became the symbol of the new social categories emerging in rapidly urbanizing France. Moreover, managers were perceived not only as representatives of a dynamic and modernizing country (contrary to the image held by the old "*petty bourgeoisie*"), but also as symbols of social success, embodying ideals of upscale social mobility. The "*Grandes écoles*" (engineering schools, Sciences Po, and ENA for the bureaucratic and political elites, plus top business schools) have for a long time played a central role in the formation of the French elites and managers (Bourdieu, 1989).

This situation sharply contrasts with the Italian context, where no specific educational set of institutions provides training for the managerial positions and the elite. Far more relevant than educational backgrounds is the accumulation of relational capital, acquired through access to the right networks (especially family ones). This explains why, up until very recently, only half of Italian managers held a master's degree (Rovati, 1991). However, managers in the Italian context do not reach a significant level of autonomy from entrepreneurs and/or capital owners (Derossi, 1978; Talamo, 1979), developing mostly at their shadows.

The case of Spain lies somehow in between the two previous countries in relation to the emergence of a managerial class. The liberalization of the economy that followed the death of Franco in 1975, and Spain's entrance in the European Economic Community in 1986, implied a radical change of the socio-economic structure (Caballero, 2003). The training and education of managers reflected the increasing internationalization of the Spanish economy, and despite the fact that holding the right relational capital continued to play a key role in accessing managerial roles, academic qualifications (notably MBAs in prestigious universities or business schools) became a key factor in selection processes in the private sector.

In the three countries the total size of managers grew over the last decades (to represent about 8% of the labor force in 2008), although this increasing trend was significantly reversed with the economic crisis of 2008, in particular in Italy and Spain as shown by Table 1..

Given the different national patterns and relative weights in the labor markets, it is quite evident that even though managers roughly share the same position within their specific national labor market, their features vary quite substantially in terms of their educational trajectory and working conditions. This is the reason why we decided to homogenize as much as possible our sample in precisely these two dimensions (education and working conditions) by selecting managers with at least a master degree, and a relative degree of autonomy at work, namely the capacity to manage their time, workload and responsibilities (i.e. those coordinating a team, or deciding upon the careers and salaries of other workers).

Table 1. Percentage of managers on the labor market by country.

	2008	2016
EU-15	9,0	6,3
Spain	7,7	4,1
France	8,4	7,3
Italy	8,3	3,9

Source: Employment by occupation and economic activity (from 2008 onwards, NACE Rev. 2) – 1 000 (lfsa_eisn2)
http://ec.europa.eu/eurostat/data/database

The spatial concentration of managers within the three countries reveals clearly distinct patterns that informed the choice of cities in this research. In France, the "*Ile de France*" region, and the city of Paris most notably, concentrate a very large number of senior professionals and managers, both in the public and private sectors. In Italy, Lombardy has the highest percentage of "*quadri and dirigenti*", with Milan coming top regarding the presence of managers employed in private companies. In the case of Spain, the metropolitan regions of Madrid and Barcelona include a large share of "*cuadros*". The choice of cities was therefore guided by the concentration of our target group. A second argument informed this choice; the study of dynamic European cities from the South West corner of the continent that are not often mentioned in the literature but are part of the backbone of the EU and contribute to the understanding of contemporary dynamics of change in the European context beyond the traditional attention paid to London and Paris.

The research was implemented in four neighborhoods in each city, with thirty interviews conducted in each neighborhood. The choice of neighborhoods was based on two basic criteria: location, and social composition of the population. In each metropolitan context, we thus identified two neighborhoods with the highest concentration of managers, and two with a more mixed social composition (although with a significant presence of managers as well), one in the city centre and another in suburbs, respectively.[3] The classic centre-suburb dichotomy raises the question of whether we can identify what Lockwood called "the urban seeking", versus the "urban fleeing" middle classes (Lockwood, 1995).

1.2. *Finding and selecting managers*

A census of the upper-middle class managers living in each city was obviously not available when generating our sample, so avoiding the potential biases introduced by the "snowball effect" (choosing respondents who are closely connected with one another) became an important concern during the process of informant's selection. The emphasis on social networks given to the questionnaire required avoiding interviewing people who knew each other well, and who had very similar and interconnected social practices and/or friends. In each city and neighborhood, a different set of entry points was used to contact managers who met the selection criteria, while trying to avoid the over-determination that could derive from selecting informants among very close circles of acquaintances, friends and/or relatives.

Even though we were not aiming at building a statistically representative sample, we tried to generate as mixed a sample of managers as possible, in terms of gender, employment in the public or private sector, generation, etc., in addition to the previously mentioned centrality/periphery and homogeneity/mixed dimensions.

In collaboration with prestigious higher education institutions (likely to have trained upper-middle class managers fitting the profile included in our research), letters were sent to alumni living in the selected neighborhoods, explaining the aim of the research, and asking for their collaboration. Professional associations were contacted as well, and many of them collaborated with us by contacting some of their members fitting the required profiles, and asking them to volunteer to be interviewed. In some cases, primary and low secondary public and private schools in the neighborhoods included in our research were also approached, asking them to help us reach the student's parents to ask for their participation. The door to door technique was applied in some cities as well, mainly in the mixed-central, and homogeneous-suburban areas. On top of all of this, we used all sorts of formal and informal networks to complete the samples of thirty managers in each of the neighborhoods.

As it is often the case in any research with a strong qualitative empirical component, finding informants was not easy, task that was made even more difficult by the busy agendas of the profiles of informants we were aiming at interviewing. Nevertheless, once the interview started, managers were generally friendly and interested in the topics included in the research, and provided very interesting insights into the nature of their professional, personal and residential trajectories, as well as about their social networks, values and strategies, that is to say, about their lives.

2. Upper-middle class residential strategies shaping the city

In previous phases of capitalist development, bourgeois groups led the process of reshaping the city, adapting it to their own expectations and interests. Research on European cities has traditionally stressed the making of bourgeois neighborhoods, and the long term influence and investment of upper social strata in city centers or in residential suburbs (Benevolo, 1993; Berengo, 1999; Pinol, 2003). With some exceptions (i.e. industrial cities, in particular), European cities have not traditionally been dominated by working classes, who were generally sent away to the periphery. In comparative terms, and with a large diversity within them, continental European cities were characterized by relatively lower levels of inequalities (thanks to welfare state redistribution), but significant levels of socio-spatial segregation, partly explained by the strategies of the middle classes (Fujita & Maloutas, 2012; Préteceille, 2007).

Evidence of upper-middle classes leaving city centers and settling in new suburban developments composed of individual detached and semi-detached houses has certainly been collected across Europe. These groups are supposed to choose lower-density areas, with a more socially and aesthetically homogeneous urban fabric, insulated from other social groups, where they may be and feel "among equals". Nevertheless, the evolution of European cities is only very partially explained by declining urban centers, massive suburbanization, overwhelming gentrification, and/or middle class flight (Phelps, Parsons, Ballas, & Dowling, 2006). US experiences of segregation and secession of the wealthiest groups (urban sprawl, strong segmentation by ethnicity and/or wealth, "gated communities") do not automatically apply to account for European urban upper-middle classes' residential strategies.

In the literatures on "suburbanization" and "gentrification" (Lees, Slater, & Wyly, 2008), upper-middle classes are identified as "segregators", social actors aiming to distance themselves from lower segments of society, looking for the benefits of the cultural and relational capital accumulation that could be derived from living among "peers" of a similar socio-economic position and status. However, in his analysis of social change in East London, Hamnett (2003) points out that "gentrification" is more about replacing a dying-out working class, and much less about displacing it.

Even with no displacement, the processes of urban transformation are conceptualized as dependent on the residential mobility of the "gentrifiers". Residential mobility is supposed to constitute the main process through which change in local social profiles is produced in contemporary metropolis. As Maloutas (2004) emphasizes, residential mobility is not the only process leading to urban change, since the social mobility of long-term residents also has an impact on processes of socio-spatial differentiation. Social mobility does not need to lead to residential out-mobility, and the socially mobile who remain in the same neighborhood where they were born and/or raised constitute an important element to be considered in the analysis of segregation trends. Following on this idea, Maloutas developed the concept of "endogenous social mobility" (or "spatially entrapped social mobility"), to refer to intra neighborhood mobility (moving within the same mixed area to maintain family and social links, and often in social order by moving from renting to home ownership), as well as to strategies of relative social avoidance.

Urban residents make choices in their experience of residence through their differential approach to the dimensions of privacy (in a spectrum ranging from anonymity to intimacy), and access to services (from public to private provision). It is particularly these spectrums that have changed over time, and over which people exert far more choice than in the past. Butler (2007) uses the metaphor of the "social tectonic" functioning of the city according to which different social classes cohabit in the same neighborhoods and pass each other by, with little or no interaction.

2.1 *Between inherited and selective belonging*

Within the debate on social classes, stratification and social status, and with some significant exceptions (Bagnasco & Negri, 1994; Maldonado, 1997; Blokland & Savage, 2001; Savage et al., 2005), little attention has been paid to the spatial dimension and to its meaning in relation to the identity that may draw from living in a certain place, whether this is the city and/or the neighborhood. Living in an urban area or in a rural one, in a large metropolitan area or in a middle size city, in a market or consumption orientated city, makes an important difference with regards to the position a person occupies in the social structure, and for the understanding of both collective actor's and individual's strategies. Recent research has analyzed the spatial dimension of class at a European scale (Recchi & Favell, 2009; Díez Medrano, 2011), while numerous studies on gentrification and suburbanization in the Anglo-Saxon world (particularly in the UK, USA, and Australia) have traditionally identified socio-spatial dynamics of gentrification, exclusion or inequalities (Ley, 1996; Smith, 1996). Pinçon and Pinçon-Charlot's work on France (1989, 2000) identified the "spatial stamp" of the bourgeoisie: a way of building and organizing "good districts" in cities, especially the largest ones. The mutual reinforcement in spatial terms enabled this particular social class to deploy effective inheritance and reproduction strategies. As these authors point out: "Spatial segregation, pushed to the extreme, is in fact an aggregation, the choice of a social group, of a class, through which it is expressing its awareness of the group's deep community of interests" (2000: 54).

Van Zanten (2009) and Oberti (2008) in France, as well as Butler and Robson (2003) and Vincent and Ball in the UK (2007), stressed the role of school choice for upper-middle class parents, and the long term influence of education in the reproduction of inequalities as central issues when analyzing social stratification patterns in contemporary societies and cities.

The notion of "elective belonging" developed by Savage et al. (2005) appears as a way to emphasize the spatial dimension of class identity formation. As Savage suggests, "elective belonging" strategies play a very important role in the process of "social location" for social classes, something quite significant in a context in which the cultural implications of occupational class have become significantly eroded for most groups. This author uses residence and cultural choices as the uppermost marks of class distinction in the Bourdieusian tradition. Unlike Savage though, we think that the position within the labor market continues to play an important role in social stratification dynamics, so we combined the study of the occupation and working career of our informants with a close look at their spatial dynamics.

The list of factors that influence the process of selecting an area to live in is potentially very long. The combination of those elements may vary as much as the specific preferences, characteristics, and living circumstances of a person. The analysis of the logic underlying the residential strategies of our informants reveals that family continues to play a central role in the life and choices of managers. Residential choice has proved very strongly influenced, if not determined, by proximity and/or convenience to maintain active family ties. This influence of the family in the residential strategies of the upper-middle classes takes different forms: inheriting a family house (by those of a more bourgeois origin), choosing to live close to their parents (or siblings) already living in an area, or going to live in a neighborhood where one of the spouses spent his/her childhood.

The role of friends also emerges quite strongly in the analysis of the residential choices of upper-middle classes. Being close to one part of the family or to friends, a "personal trajectory factor" following Martin-Brelot, Grossetti, Eckert, Gritsai, and Kovács (2009) terminology, emerges as a very strong driving factor in the choice of neighborhood. This constitutes an element of reproduction of social inequalities (inheritance) and stability within the territory, to a large extent preventing mobility towards other areas of the city that may, potentially, be considered more attractive with respect to other factors (location, public services, social composition, etc.) for those who can afford it. This factor anchors some upper-middle classes in more mixed and peripheral areas of the city, where they (or their partners) were born and/or raised, and where their family resides, despite their drives to move towards more exclusive, homogeneous and/or central neighborhoods. This process, related to Malouta's concept of "endogenous social mobility" (Maloutas, 2004), proved crucial for the choice of neighborhood among our informants, evidence of managers "rootedness".

2.2. *No extreme segregation*

A significant part of the social status of upper-middle class managers is directly related to their residential trajectories and strategies. European cities are in part socially mix. This does not mean so much cohesiveness, as cohabitating spaces with often limited interaction between different groups who live relatively separated lives (Davidson, 2010). Classic research on the spatial proximity of social groups had already identified strategies of social differentiation and distance (Chamboredon & Lemaire, 1970; DeFilippis & Fraser, 2008), defining concepts such as "micro-segregation" (coexistence of very polarized groups in small spaces, with very high levels of segregation), and "micro-fragmentation" (difficult coexistence and mutual obliviousness of very different social groups living together in a small space). Blokland (2003), in her comparative research about two neighborhoods in the US and in the Netherlands, also showed how location does not by itself produce community, and at most 'communities make use of the location to varying degrees'.

If proximity does not necessarily mean cohesion, segregation does not need to mean isolation either. Elaborating on this idea, Maloutas (2004) criticizes the "spatial reductionism" of the Chicago school, and a good part of urban sociology thereafter, which equals social mobility (through residential mobility of the upwardly mobile) with urban segregation. Segregation is seen as the result of the departure of specific social groups, and/or as a consequence of the residential mobility of incoming residents with a different profile to the long-term residents of an area. For instance, moving to a gated community (the most "insulated" form of separation from other social groups) does not necessarily mean that the upper-middle classes are increasing their level of segregation either, since the fact that they used to live in closer proximity to lower classes did not mean that they were sharing much with them to start with, thus emphasizing the need to look beyond the presence or absence of interaction between social groups, to focus on the "quality" and "nature" of the existing interactions (Álvarez-Rivadulla, 2007).

In our research we analyzed those strategies by reviewing our empirical evidence on how the managers of the sample selected their city and neighborhood of residence, where they used to live before, their feelings about their neighborhood, and the metropolitan area they live in, as well as their use of services at the different territorial scales, allowing us to analyze their degree of "rootedness" in the urban fabric. In general terms, there is relatively little evidence of structural and systematic "privatism", or retreat from the public sphere, among the upper-middle class managers of our research. This does not mean that they are ready to mix with lower social groups all the time and in all dimensions.

3. Accounting for "partial exit" strategies and territorial "rootedness"

To better grasp whether European managers were becoming "deterritorialized", abandoning public space, and avoiding interactions with other social groups, we used Hirschman's (1970) classic "exit, voice and loyalty" conceptual framework. Applied to the socio-political institutional logic of a world order organized around nation-states, Hirschman's model implies that if individuals cannot achieve their goals while living in their home country, they may choose to "exit" and leave for another country in order either to escape constraints, or to find a job, for instance.

The European integration process greatly facilitates the development of strategies of internationalization of the European upper-middle classes (Díez-Medrano, 2010; Recchi & Favell, 2009). These groups take advantage of the opportunities associated to the concept of European citizenship to acquire and expand different forms of capital which they can later cash in their national labor markets as an element of differentiation. In practice, though, the "exit" option remains relatively rare in Europe, for as Favell (2008) has shown, it often involves considerable personal costs.

Faced with the complexity and relative difficulties associated to complete "exit" through emigration to another country, individuals may choose to develop "partial exit" strategies from their national societies (i.e. to "exit" in order to come back, or to "exit" in some domains but not others). This "partial exit" may take the form of a limited "exit" to a foreign country for a certain period of time to accumulate experience, skills and/or resources (foreign languages, degrees, professional experience or qualifications, etc.) before going back to their home country. It may also refer to regular trips to foreign countries for professional or personal reasons to strengthen networks abroad. "Partial exit" may indicate that some dimensions of their personal or social life are transnationalized: owning real estate in a foreign country, the educational trajectory of their children, or avoiding public services in their hometown (i.e. health services).

To analyze the "exit" and/or "partial exit" strategies of our respondents with regards to the international dimension, we built a series of indicators based on the professional and educational experiences of our informants, their friendship networks abroad, the frequency of their international travelling (both for professional and personal reasons), the foreign cities they consider to know well, their exposure to international media outlets, and the priority they attach to the international dimension in the education of their children.

Our upper-middle class informants showed clear signs of being "partially exiting" their national societies in many of those indicators, obviously with different combinations of strategies and intensities in the different cities, thus validating the adequacy of those indicators to apprehend the development of this dimension in the lives of European urban managers. They are not the only social group that may be including this transnational dimension within their lives. We could notably think of populations of immigrant origin as another group strongly exposed to dynamics of building identities and practices in different layers and levels at the same time (Kennedy & Roudometof, 2002; Mau & Mewes, 2009), but the relatively privileged position of our managers puts them in a particularly favorable context to take advantage of the opportunities that the growing internationalization of our societies create to promote their own personal and professional trajectories (Mau, 2010; Smith, 2002).

At the urban level, "exit" strategies may indicate living in relative isolation, complete segregation or absolute secession (e.g. gated communities), without investing in the urban context, while "partial exit" implies diversifying spheres of interaction with the objective of maximizing their interests, and minimizing (the risk of) potential costs. To understand whether "exit" or "partial exit" strategies at the urban level were put into practice by our respondents, we considered four dimensions: the residential strategies mentioned earlier; the frequency and use of public services, both at the neighborhood, and city level (school, transportation, postal office, health services, local libraries, sports facilities); consumption practices, considering the frequency and use of cultural spaces (cinemas and theatres), and everyday life (bars, restaurants, shops); as well as sociability patterns (family, friendships, and neighborhood ties).

Our research shows that social dynamics in the urban space are generally more complex than extreme mutual avoidance (isolation and segregation), or the colonization of neighborhoods (gentrification). Upper-middle class managers can choose to "exit" in some dimensions, for instance sending their children to international schools or universities, and investing in others (consumption, friendships, etc.). In order to study "rootedness" we used the following indicators: being born and living in the same neighbourhood/city, length of living in the same neighbourhood/city; use of services in the neighbourhood/city; use of public services (in particular schools, health system and transportation); engagement in the neighbourhood/city; presence of family and friends in the neighbourhood/city. The mix of these answers give rise to our understanding of "rootedness" and strategies of exit/partial exit. These indicators are independent from the fact of living in a gated community, or a mixed central neighbourhood. It is only criss-crossing these indicators that we can say whether our managers "exit" or "partial exit" and at what extent they are territorialized.

Most of the times upper-middle class managers skillfully combine proximity and distance in relation to other social groups, selecting the dimensions they are willing to share with other social groups, and to find a more segregated social environment for themselves and their families. Control is everything here, a clear case of power relations. The social composition of the neighborhood seems crucial in this respect: the more managers live in a residential suburb, or in an exclusive city centre neighborhood, the more relaxed they seem to be about encountering people from other social and ethnic groups at the supermarket, the public transportation system, the swimming pool, the school, or the park. By contrast, the more they reside in a mixed neighborhood, the more precise they become in trying to control the conditions under which social mixing takes place.

They search to distance themselves in certain domains, but less so in others, by developing strategies that allow them to select the public sphere they are willing to share with lower classes. These strategies of distance and proximity imply that upper-middle classes often choose to remain "rooted" in their neighborhoods and cities through their interaction with their own social networks, a selective use of public services, and/or their presence in specific public spaces.

The use of public services in those mixed contexts reveals the capacity of these groups to articulate the combinations that allow them to feel comfortably in control of the degree of social mixing they are exposed to. It is mainly thanks to this capacity that most of the interviewees see their cities as resource-rich environments in terms of services and networks of friends and families allowing them to lead successful professional careers, while raising a family and having a dynamic social life. In this sense, we only partially agree with Sennett (2002) who when mentioning elites, wrote that: "At the top of this new capitalist system, the global inhabitants of modern cities tend to use the city as a space for economic activity without any desire to participate as citizens (…). Indifference about physical spaces, I could be here or anywhere (…), a regime of differences that are non- interactive (…), you get a regime of geographical, educational and even to some extent leisure segregation in which class, race and ethnic differences are managed in the city by principles of non interaction (…). Difference produces indifference" (Sennett, 2002: 45–46).

Forms of "elective" or "selective belonging" are crucial for managers, their narratives clearly reveal why they live in a certain place (city or neighborhood), showing equally well "that a certain place" could not be anywhere else, because it is full of social meaning for them due to family ties, friendship, and places (schools, services, leisure) they know well and appreciate. By contrast, what we called the control of distance and proximity appears to be similar to what Sennett argues in relation to the managing of differences. Managers deploy their resources to use distance and proximity according to their norms, expectations and/or interests. Relative and limited social mixing under strict control describes the strategy they follow by them when balancing proximity and distance in relation to other social groups, using their resources to maximize their own interests. The narratives of managers stress that differences result in "non-interactions", which do not signal indifference, rather express avoidance, a precise strategy within power relations.

4. Conclusion

Urban upper-middle classes proved to be central actors in shaping their cities, and in the making of their nation societies. The rise of mobilities in recent times paved the way for important changes within cities and societies, with an increased influence of upper-middle classes who can mobilize important economic, human and social resources to "exit" from their cities and national societies, disinvesting on them, while still taking advantage of a process of "deterritorialization" with important consequences at the collective level.

Investigating the extent to which upper-middle classes shaped (and are still shaping) cities through their residential strategies and practices is a challenge for many reasons, even more so when international comparison is considered. First, it forces to better define and understand who the upper-middle classes are, taking into account the local and national patterns of social groups composing them, their similarities and differences. The first part of this article aimed at providing insights on the different social meaning that upper-middle classes, and the social groups composing them, can have considering the knowledge acquired in the three Western European countries and four cities studied in this project. Investigating residential strategies and every day practices requires both attention to statistical data, and original empirical fieldwork. In this contribution we highlighted how we generated our data through personal interviews with European managers, as well as the conceptual framework that informed the understanding of "rootedness", "exit" or "partial exit" strategies developed by our managers.

On the whole, our results tell us that these managers remain locally "rooted" while skillfully playing a complex game of distance and proximity with regard to other social groups in their urban environments. But they also invest resources in their cities and neighborhoods, and only look for "urban exit" strategies in very specific contexts. The managers interviewed in this research remain in control of the social and spatial distance they keep with diverse social and ethnic groups living in their areas of residence. These upper-middle classes have therefore a "globalized" mind but remain firmly "rooted" in their local urban environments, where they belong to relatively dense networks of friends and family, and where they invest in the functioning of the local social and political sphere.

The research shows how difficult and fruitful is to analyze with different levels of social organizations comparing them: we included 16 neighborhoods in four cities from three different countries in our comparison. Playing with the different scales and the comparison has proven hard to manage (in three languages) but has given a profound depth to the material we collected, and to the analysis. But it takes time. Also, the convenient sample we selected allowed us to use our 480 face to face interviews to combine closed questions (allowing for some quantitative analysis), open questions (qualitative interpretation), and some measures of sociability. The combination of those types of data and methods of analysis proved crucial to provide key interpretations to these complex phenomena. Some social groups become very local, but some are in the making at the urban or the European scale. The comparison has provided evidence on the slow but gradual making of something resembling a European urban upper-middle class.

Within an ideal world, with no limitations of time and money, we would have liked to combine this research with a general survey at the national level based on a representative sample, and combined with more systematic ethnographic survey of the different neighborhoods. This was beyond our possibility.

The questionnaire, and the type of operationalization of research data we used, proved quite fruitful, and it is going to be used in different research projects. One of those projects is a study of Italian, Spanish and French managers in Hong Kong using the same questionnaire. A second one aims at working with French residents in London, and English residents in Paris.

The intensely empirical, and largely experimental nature of this research project on urban upper-middle classes can provide interesting insights for the study of urban elites. The methodological and conceptual challenges we confronted when designing and putting this project in practice could provide some guidance on what we consider to be a fruitful field of research to understand the future evolution of cities, and the spatial/mobility dimension of class making.

Notes

1. It is worth reminding that despite all claims about increasing mobility, Europeans who decide to move their residence abroad are less than 3% each year.
2. In this contribution we do not consider the transnational dimension of this phenomenon, which has been treated elsewhere (Andreotti, Le Galès, and Moreno-Fuentes, 2012 and Andreotti et al., 2014).
3. The selection of neighbourhoods in Paris was based on the remarkably precise analysis and classification carried out by Preteceille (2006) on the social composition of the small statistical units comprising neighbourhoods in the city. In the Milan case, the selection was based on the ecological analysis of Pratschke (2007).

Disclosure statement

No potential conflict of interest was reported by the authors.

ORCID

Francisco Javier Moreno-Fuentes http://orcid.org/0000-0002-1393-6225

References

Álvarez-Rivadulla, Maria José. (2007). Environment and Planning A: Economy and Space. *39*(1), 47–63

Andreotti, Alberta, Le Galès, Patrick, & Moreno-Fuentes, Francisco Javier. (2014). *Globalized minds, roots in the city. Urban upper-middle classes in Europe*. Oxford: Wiley Blackwell.

Atkinson, Antony B. (2015). *Inequality: What can be done?* Cambridge (Mass): Harvard University Press.

Atkinson, Roland, & Gary, Bridge (Eds.). (2005). *Gentrification in a global context: The new urban colonialism*. London: Routledge.

Atkinson, Roland, Parker, Simon, & Burrows, Roger. (2017). Elite Formation, Power and Space in Contemporary London. *Theory, Culture & Society, 34*(5–6), 179–200.

Bagnasco, Arnaldo, & Le Galès, Patrick (2000). *Cities in contemporary Europe*. Cambridge: Cambridge University Press.

Bagnasco, Arnaldo, & Negri, Nicola. (1994). *Classi, ceti, persone. Esercizi di analisi sociale localizzata*. Liguori: Napoli.

Bartolini, Stefano. (2005). *Restructuring Europe. Centre formation, system building and political structuring between the nation state and the EU*. Oxford: Oxford University Press.

Beaverstock, Jonathan V, Hall, Sarah, & Wainwright, Thomas. (2011). Servicing the super-rich: new financial elites and the rise of the private wealth management retail ecology,regional. *Studies,47:6,834-849*.

Benevolo, Leonardo. (1993). *The European City*. Oxford: Basil Blackwell.

Berengo, Marino. (1999). *L'Europa delle città. Il voloto della società urbana europea tra Medioevo ed Età Moderna*. Turin: Einaudi.

Blokland, Talja. (2003). *Urban Bonds*. Cambridge: Polity Press.

Blokland, Talja, & Savage, Mike. (2001). Networks, class and place. *International Journal of Urban and Regional Research, 25*, 221–226.

Bourdieu, Pierre. (1989). *La Noblesse d'Etat. Grandes écoles et esprit de corps*. Paris: Les Editions de Minuit.

Burawoy, Michael. (2005). For public sociology. *American Sociological Review, 70*(1), 4–28.

Burrows, Roger, Webber, Richard, & Atkinson, Roland. (2017). Welcome to 'Pikettyville'? Mapping London's alpha territories. *The Sociological Review, 65*(2), 184–201.

Butler, Tim. (2007). For Gentrification? *Environment and Planning A, 39*, 162–181.

Butler, Tim, & Robson, Gary. (2003). Plotting the middle classes: Gentrification and circuits of education in London. *Housing Studies, 18*(1), 5–28.

Caballero, Miguez Gonzalo. (2003). *The evolution of institutions and state governing public choice in the second half of twentieth-century Spain*. Santiago de Compostela: Universidade de Santiago de Compostela.

Chamboredon, J-C., & Lemaire, Madeleine. (1970). Proximite spatiale et distance sociale. Les grands ensembles et leur peuplement. *Revue Française De Sociologie, 11*, 3–33.

Crouch, Colin. (1999). *Social changes in Western Europe*. Oxford: Oxford University Press.

Davidson, Mark. (2010). Love thy neighbour? Social mixing in London's gentrification frontiers. *Environment and Planning A, 42*(3), 524–544.

Davis, Aeron, & Williams, Karel. (2017). Introduction: Elites and power after financialization. *Theory, Culture & Society, 34*(5–6), 3–26.

DeFilippis, James, & Fraser, James. (2008). *Why do we want mixed-income housing and neighborhoods? ESRC seminar: Gentrification and social mixing*. Kings College London. http://muse.jhu.edu/books/9781438433073. Downloaded 12-08-2014.

Derossi, Flavia. (1978). *L'illusione tecnocratica. Il potere dei dirigenti nell'industria italiana*. Milano: Etas Libri.

Díez-Medrano, Juan (2010 April). *A new society in the making: European* integration *and* european social groups. KFG Working Paper Series, No. 12, Kolleg-Forschergruppe (KFG) "The Transformative Power of Europe", Freie Universität Berlin.

Elliott, Anthony, & Urry, John. (2010). *Mobile lives*. New York: Routledge.

Esping-Andersen, Gosta. (1990). *The three worlds of welfare capitalism*. Princeton: Princeton University Press.

Eurostat. (2015). *Air transports statistics*. Brussels, EU: Author. http://ec.europa.eu/eurostat/statistics-explained/index.php/Air_transport_statistics.

Favell, Adrian. (2008). *Eurostars and Eurocities*. Oxford: Blackwell.

Favell, Adrian, & Guiraudon, Virginie (Eds.). (2011). *Sociology of the European Union*. London: Palgrave.

Favell, Adrian, & Recchi, Ettore. (2011). Social mobility and spatial mobility. In Adrian Favell & Virginie Guiraudon (eds), *Sociology of the European Union* (pp. 50–75). London: Palgrave.

Fligstein, Neil. (2008). *Euroclash: The EU, European identity, and the future of Europe*. Oxford: Oxford University Press.
Fujita, Kuniko, & Maloutas, Thomas (Eds.). (2012). *Residential segregation in comparative perspective: Making sense of contextual diversity*. Farnham: Ashgate.
Geographical and labour market mobility. Report No. 3. Dublin: European Foundation for the Improvement of Living and Working Conditions.
Gouldner, Alvin. (1979). *The future of intellectuals and the rise of the new class*. New York: The Seabury Press.
Gusftason, Peter. (2009). More cosmopolitan, no less local. *European Societies, 11*(1), 25–47.
Hamnett, Chris. (2003). *Unequal city: London in the global arena*. New York: Routledge.
Hannerz, Ulf. (2003). *Transnational connections: Culture, people, places*. London: Routledge.
Hirschman, Albert. (1970). *Exit, voice, and loyalty: Responses to decline in firms, organizations, and states*. Cambridge: Harvard University Press.
Jansson, André. (2016). Mobile elites: Understanding the ambiguous lifeworlds of Sojourners, Dwellers and Homecomers. *European Journal of Cultural Studies, 19*(5), 421–434.
Kaelble, Hartmut. (1990). *A social history of western Europe, 1880-1980*. Dublin: Gill & Macmillan.
Kaufman, Vincent. (2003). *Rethinking mobility*. London: Routledge.
Kennedy, Paul. (2007). Global transformations but local "bubble" lives: Taking a reality check on some globalization concepts. *Globalizations, 4*(2), 267–283.
Kennedy, Paul, & Roudometof, Victor (Eds.). (2002). *Communities across borders under globalising conditions: New immigrants and transnational cultures*. New York, London: Routledge.
Larsen, Jonas, Urry, John, & Axhausen, Kay. (2006). *Mobilities, networks, geographies*. Aldershot: Ashgate.
Lees, Loretta, Slater, Tom, & Wyly, Elvin. (2008). *Gentrification*. New York: Routledge.
Ley, Dey. (1996). *The new middle class and the remaking of the central city*. New York, NY: Oxford University Press.
Lockwood, David. (1995). Introduction: Making out the middle class(es). In Tim Butler & Mike Savage (eds), *Social change and the middle classes* (pp. 1–14). Londres: UCL Press.
Maldonado, Leal Jesús. (1997). Sociología del espacio: El orden espacial de las relaciones sociales. *Política Y Sociedad, 25*, 21–36.
Maloutas, Thomas. (2004). Segregation and residential mobility: Spatially entrapped social mobility and its impact on segregation in Athens. *European Urban and Regional Studies, 11* (3), 195–211.
Mann, Michael. (2013). *The sources of social Power, vol 3, Global empires and revolution, 1890–1945*. Cambridge: Cambridge University Press.
Martin-Brelot, Hélène, Grossetti, Michel, Eckert, Denis, Gritsai, Olga, & Kovács, Zoltan. (2009). Not so mobile 'creative class': A European perspective. *GaWC Research Bulletin*, 306.
Mau, Stephen. (2010). *Social transnationalism: Lifeworlds beyond the Nation State*. London, New York: Routledge.
Mau, Stephen, & Mewes, Jan. (2009). Class divided within transnationalisation. The German population and its cross-border practices. In Timo Ohnmacht, Hanja Maksim, & Max Bergman (Eds.), *Mobility and inequality* (pp. 165–186). Abingdon: Ashgate.
Medrano, Díez medrano, Jude. (2011). Europeanisation and class. In Favell, Adrian & Guiraudon Virginie (Eds.), *Sociology of european union* (pp. 28-49). Basingstoke: Palgrave.
Mendras, Henri. (1997). *L'Europe des Européens: Sociologie de l'Europe occidentale*. Paris: Gallimard.
Oberti, Marco. (2008). Ségrégation urbaine et scolaire. In Agnès Van Zanten (Eds.), *Dictionnaire de l'éducation* (pp. 619–621). Paris: Presses Universitaires de France.
Phelps, Nicholas, Parsons, Nick, Ballas, Dimitris, & Dowling, Andrew. (2006). *Post-suburban Europe: Planning and politics at the margins of Europe's Capital Cities*. Basingstoke: Palgrave-Macmillan.
Piketty, Thomas. (2014). *Capital in the twenty-first century*. Cambridge: Harvard University Press.
Pinol, Jean Luc. (2003). *Histoire d'Europe Urbaine*. Paris: Presses Universitaires de France.

Pratschke, Jonathan. (2007). L'articolazione territoriale dello svantaggio sociale in Italia: Una nuova misura multidimensionale per piccole aree geografiche. In Andrea Brandolini & Chiara Saraceno (Eds.), *Povertà e Benessere. Una geografia delle disuguaglianze in Italia* (pp. 113–136). Bologne: Il Mulino.

Préteceille, Édmond. (2007). Is gentrification a useful paradigm to analyze social changes in the Paris metropolis? *Environment and Planning A*, *39*, 10–31.

Recchi, Ettore. (2015). *Mobile Europe: The theory and practice of free movement in the EU*. Basingstoke: Palgrave-Macmillan.

Recchi, Ettore, & Favell, Adrian. (2009). *Pioneers of European integration: Citizenship and mobility in the EU*. Cheltenham: Edward Elgar.

Robinson, William, & Harris, Jerry. (2000). Towards a global ruling class? Globalization and the transnational capitalist class. *Science and Society*, *64*(1), 11–54.

Rovati, Giancarlo. (1991). *Un ritratto dei dirigenti italiani*. Torino.:Edizioni della Fondazione Giovanni Agnelli.

Savage, Mike, Bagnall, Gaynor, & Longhurst, Brian. (2005). *Globalization and Belonging*. London: Sage.

Savage, Mike, & Williams, Karel (Eds.). (2008). *Remembering elites*. Oxford: Wiley Blackwell.

Sennett, Richard. (2002). Cosmopolitanism and the social experience of cities. In Steven Vertovec & Cohen Robin (Eds.), *Conceiving cosmopolitanism: Theory, context and practice* (pp. 42–47). Oxford: Oxford University Press.

Sklair, Leslie. (2001). *The transnational capitalist class*. Oxford: Blackwell.

Sklair, Leslie. (2005). The transnational capitalist class and contemporary architecture in globalizing cities. *International Journal of Urban and Regional Research*, *29*(3), 485–500.

Smith, Neil. (1996). *The new urban frontier: Gentrification and the revanchist city*. London: Routledge.

Smith, Neil. (2002). New globalism, new urbanism: Gentrification as global urban strategy. *Antipode*, *34*(3), 427–450.

Talamo, Magda. (1979). *I dirigenti industriali in Italia. Autorità, comando e responsabilità sociali*. Torino: Einaudi.

Therborn, Goran. (1995). *European modernity and beyond*. London: Sage.

Urry, John. (2000). *Sociology beyond societies*. London: Routledge.

Van Zanten, Agnès. (2009). *Choisir son école. Stratégies familiales et médiations locales*. Paris: Presses Universitaires de France.

Vandenbrande, Tom (Ed.). (2006). *Mobility in Europe. Analysis of the 2005 Eurobameter Survey on*.

Zunz, Olivier, Schoppa, Leonard, & Hiwatari, Nobuhiro (Eds.). (2002). *Social contracts under stress: The middle classes of America, Europe, and Japan at the turn of the century*. New York: The Russell Sage Foundation.

Naked elites: unveiling embodied markers of superiority through co-performance ethnography in gentrified Brooklyn's Park Slope

Lidia Katia Consiglia Manzo

ABSTRACT

In the super-gentrified neighborhood of Brooklyn's Park Slope, elites joined democratic associations, such as karate dojos or food coops, to interact with and become socially accepted by the local community. Co-performance ethnography made it possible to recognize that progressive moral values led instead to exclusive behaviors. Drawing on symbolic interactionism and socio-cultural analysis, this article provides a definition of performance as the paradigm-driven methodological tool to unveil the hidden, micro-strategies of elite reproduction. I analyze how social mixing in gentrified neighborhoods encourages competing identities to reproduce symbolic boundaries, ultimately serving as a basis for socio-spatial inequality and conflict. The production of identities, behavior, and emotions highlights elites performance of a contextualized version of "self" to assert their superior status. It also shows how these relations of domination shape elites' lives. These points call attention to the connections between capital accumulation and the hermeneutical agency of the body.

Introduction

Key global processes of gentrification (Lees, Shin, & Lopez-Morales, 2015, 2016) and financialization (Aalbers, 2016; Sassen, 2014) have been a catalyst to urban research in the exclusionary effects of the growing domination of affluent elites in the contemporary city (Bridge, Butler, & Lees, 2012; Butler & Lees, 2006; Lees, 2008, 2013; Lees & Ferreri, 2016; Rose et al., 2013; Slater, 2006; Zukin, Kasinitz, & Chen, 2015). In the United States, the concentration of income and wealth among elites has increased dramatically since the early 1980s, peaking in the early 2000s (Atkinson & Piketty, 2010; Keister & Lee, 2014). Particularly in the last decade, the City of New York has attracted a growing population of wealthy foreigners facilitating the workings of capitalism through global financial investment (Aalbers, 2008; Fields, 2015; Krippner, 2005) that has transformed the City's real estate market (Story & Saul, 2015), as well as its socio-cultural fabric (Zukin, 1982, 2010). However, the implications of these rapid transformations on the lives of ordinary urban dwellers, especially those with limited economic resources, are poorly understood (Slater, 2009).

While there has been a proliferation of publications in recent years with respect to critical and geographical accounts of the economic elite (Bennett et al., 2009; Fernandez, Hofman, & Aalbers, 2016; Hay, 2013), these studies have not connected those macro political-economic and geographical processes with the "micrological textures of power" (Katz, 1992, p. 264) involved in the social reproduction of class inequality. This is particularly true with regard to social and economic domination of the upper classes at the community and neighborhood levels. There is little analysis or recognition of the ways they think or act in interactions with people of lower classes. Elites engage in everyday practices with these city dwellers. Unlike in other settings, they are not the ruling actors in these particular situations. For the most part, economic elites are seen as those who exude their class privilege and symbols of high social status through their conspicuous consumption preferences (Myers & Kent, 2004; Veblen, 1899) and distinct, reserved lifestyle. For example, economic elites often isolate themselves in luxurious, gated communities or exclusive, suburban country clubs (Beaverstock, Hubbard, & Short, 2004; Borsdorf & Hidalgo, 2008; Cousin & Chauvin, 2013, 2014; Pow, 2011). On the contrary, Andreotti, Le Galès, and Moreno-Fuentes (2015) found that, despite the rise of high-skilled professional global mobility, upper-class people are profoundly embedded and territorialized in their cities and neighborhoods.

Utilizing a symbolic interactionist[1] approach and drawing on Lamont's (1992) and Bourdieu's (1984) socio-cultural theories of class and distinction, this article has two goals: 1) to define a heuristic method for the investigation into how cities' elite populations produce class divisions and 2) to examine the performative dimension of elitist behavior by documenting the production of symbolic boundaries in inter-class interactions. This paper addresses the conditions under which symbolic boundaries generate inequality and exclusion, contributing to existing literature on the role of internal, embodied manifestation of superiority in the geography of urban elite.

The article presents evidence from an ethnography conducted in Brooklyn's Park Slope, one of the most desirable neighborhoods of New York City, that, in the last 40 years, has been the "frontier" (Smith, 1996) of intense investment and conspicuous consumption. Today, Park Slope is a super-gentrified urban community (Lees, 2003; Manzo, 2014; Slater, 2003) populated by economic elites that represent a sample of the top one percent earners in the United States: chief executive officers, celebrities, doctors, lawyers, media entrepreneurs, and Wall Street financiers. Two ethnographic case studies were chosen: a popular martial arts studio and a food cooperative market. These two unique places are governed by distinct social codes. For example, in a karate dojo, it does not matter if you are a "White, well-educated financier" or a "West Indian construction worker." The value of your practice, the level of your belt, and an unconditional respect for the Master are the only power forces at play. At the Food Coop, all members are equal because, in order to shop there, one must volunteer 2 hours and 45 minutes every four weeks with a group of the other 16,000 members. The socio-demographic composition of Coop members strikingly resembles that of the entire borough of Brooklyn. The karate dojo and the food coop were both critical cases as they represented the least likely places in which to find practices of elite distinction.

Ethnography helped understand that elite reproduction still takes place in these settings. When they are "nakedly exposed" in non-elite spaces, elites do not have the ability to control the system of norms and values. Thus, far more subtle micro-strategies are required to distinguish themselves from others. The notion of "off-stage" practices is central to this argument and is the key for the interpretation of signs of superiority within the upper class. These signs are unveiled in interactions where elites must "take off" their social status and, consequently, surrender their power.

To understand performance as the paradigm-driven methodological tool to explore expressive and embodied micro-elements of elite reproduction, I first summarize how this paradigm is applied to the field of urban ethnography in elite research. This summary is followed by an analysis of the conceptual apparatus that informed my field research and provide empirical evidence of how co-performance activities allow the ethnographer the opportunity to study embodied forms of elite distinction more in depth. Lastly, I discuss the complexity and ambiguity of elites' relations of domination. I also highlight how they impose their moral positions on local communities in everyday practices, challenging critical geographers to engage in symbolic interactionism to refine our knowledge of how issues of distinction, segregation, and hegemony work spatially.

The performance paradigm in urban elite ethnographies

The performative turn in ethnography has progressed beyond the recognition that social life is performative, moving toward an understanding of performance as a paradigm and a tool for conducting research anchored in dialogue, praxis, and the politics of resistance (Conquergood, 1989; Denzin, 2014; Turner & Bruner, 1986). Linked to the epistemological and political commitment of critical geography, this perspective privileges embodied forms of research centered on the ethnographer's negotiated and delicate face-to-face encounters.

Turner advocates for performance-sensitive research (1986) by pointing out the determinant role of the ethnographer in "coactivity," or as a co-performer able to understand embodied meanings "on their pulses" (1975, p. 29). If performance applies to both the researcher and the researched (Conquergood, 1991, 1992), the engagement in a "shared" fieldwork experience becomes an hermeneutical agency, a more inter-subjective sphere that enables the interpretation and transfer of knowledge (M. Jackson, 1989). According to Butler, performance is the perspective from which to delve into the embodied construction of identities "through language, gesture, and all manner of symbolic social sign" (1988, p. 519). Therefore, co-performance activities allow the researcher to gather undetermined aspects of social interaction. The researcher experiences the very same evaluations that emerge out of participating in shared experiences. These experiences can usually only be understood through the examination of how different individuals come to interpret situations distinctly, but performance ethnography allows the researcher to understand these firsthand (Hutter, 2012).

According to Thrift (2008), geographers should focus on non-representational work as a way to understand meaning as not residing in something but as generated through processes. Attention to "multiplying performative methodologies" present new concerns for urban researchers, "which allow their participants equal rights to disclosure, through dialogical actions rather than texts, through relation rather than representation" (Ibid., p.

148). Within the move to performance in geography (Gregson & Rose, 2000) and the recent interest shown in the embodied aspects of identity in situated contexts (Clifford, Cope, Gillespie, & French, 2016), the approach followed in this article can be traced to Goffman's analyses of interaction and the performing self (Goffman, 1956, 1983).

Theoretical perspectives on distinction, symbolic boundaries, and the significance of the information given when the self is "OFF"

The study of elite distinctiveness is understood in this work as the study of a performance in an evolving process, rather than a constant postulate or premise (J. Clifford, 1988; Conquergood, 1991). The creation and maintenance of distance recalls basic processes in social life that emphasize the boundaries between those included (the pure) and those excluded (the impure) across different groups (Durkheim, 1965; Simmel, 1950). Within this tradition, the tacit "symbolic boundaries," (Lamont & Fournier, 1992; Lamont & Molnár, 2002) conceptual distinctions that we make to categorize our experiences, are particularly relevant in this endeavor. Lamont argues that "[b]oundary work is an *intrinsic* part of the process of constituting the self" (1992, p. 11, italics is mine). Elites use symbolic boundaries in the process of performing their identity and class status, especially in relations with lower classes, to sustain a sense of superiority. Such a performance is used to enforce, maintain, normalize, or rationalize inter-class divides, in sum contributing to class reproduction. However, basic inferences of similarities and differences that define identity often allow an "unintended effect" to transpire.

Actual processes of embodiment of dominant class dispositions, or habitus, are indeed reiterated in practice[2] and played out accordingly to distinction, as significantly analyzed by Bourdieu (1984). Central to his thesis are the less obvious expressions of material domination – defined in this article as embodied markers of superiority. According to Auslander, social stratification appears clearly in practices embedded in a particular habitus "because class inequalities and the dominance of one class over another occur covertly. Rather than the application of overt force, symbolic power is harnessed to maintain class distinction and the appearance of their naturalness" (2008, p. 69). To account for the embodied forms of elite distinction, researchers have to observe "more subtle, but also more deeply anchored means of demonstrating social worth" (Daloz, 2010, p. 81). However, this may not be an easy feat. Elites' attitudes of superiority are not officially or too openly expressed, and they may be abstruse in interactions with those of other social classes where they may feel obligated to "level the playing field" with regard to external symbols of wealth. In this light, Bourdieu's argument underlines a heuristic method for thinking about the significance of the information given during more hidden practices, those aforementioned "unintended effects", eventually linking his view to that of theatrical production as social practice.

Goffman's dramaturgical model (1956) of the micro order of interaction is of particular interest and can be conceptually employed in the understanding of how elites convey such signs of superiority. By using the "stage" and the "theatre" as key organizing metaphors, Goffman sustains how versions of a "self" are differently played out in more or less public performances accordingly to "front" or "back" spatial regions. Particularly, the term back "refers to the area where the actor(s) are 'offstage' and can relax, let their guard down"

(Richards, 2001, p. 61). The symbolic interactionist agenda and the importance of role and self-presentation came to dominate social psychological theorizing about the self. Works on social identity theory have contributed enormously to our understanding of intergroup relations and group processes (Abrams, 1996; Abrams & Hogg, 2001; Hogg, 2001; Stets & Burke, 2000). This research has encouraged the integration of different levels of conceptualization and explanation of the performance of self. Different components of the self-structure combine to embody differing goals and a "salience hierarchy" (Stets & Burke, 2000, p. 231) is involved in the enactment of a performance in a particular context when more than one self may be appropriate. Therefore, the definition of context and the conditions that might determine it become of extreme importance when we think about the negotiation of different "self-conceptions" (Abrams, 1996, p. 145) in inter-class symbolic boundaries. This negotiating process forces elites to resolve the "distress" between the enactment of their "private" self – with cognitions on one's own attitudes, traits, feelings, "collective" self – with cognitions about group memberships and affiliations of all types, or a combination of the two.

According to Goffman (1951), close observations that consider "the pressures that play upon behavior[3]" (p. 295) in social situations provoke both expressions given ON and expressions given OFF, with the latter being "the more theatrical and contextual kind, the nonverbal, presumably unintentional kind" (Goffman, 1956, p. 16). This raises questions about the social significance of the information given OFF. Information given OFF is significant because it allows inherent signs of class habitus to transpire.

Park slope, the *gold coast* of victorian brooklynites and super-gentrifiers

The "corporatization of gentrification" in New York City reflects the city's dramatic post-industrial socio-economic changes (Lees, 2000; Smith & DeFilippis, 1999). The phenomenon has contributed to the redevelopment and rebirth of formerly industrial areas into chic residential communities, making Park Slope one of the most fashionable neighborhoods of southwestern Brooklyn (see map in Figure 1).

After the Great Depression, the "White flight" of urban dwellers led residents to the suburbs in the mid-1900s. Today, Park Slope is once again becoming a neighborhood for the rich, as it was up until the end of the 19th century, when the lawyer and railroad magnate Edwin Litchfield sold 526-acres of what eventually became Prospect Park,[4] to New York City. Park Slope was Brooklyn's Gold Coast (K. T. Jackson, 2004), a distinguished place for the extremely wealthy to develop vacant property and live. The grandiose mansions and charming Victorian brownstones built by those early elites proved to be an architectural anchor that attracted the first wave of gentrifiers to Park Slope more than a century later (Osman, 2011).

Park Slope's reputation as a hub for elites is further supported by its well-designed public places and urban landscapes. Other key place-making elements include: two lively commercial corridors (Fifth Avenue and Seventh Avenue); multimodal transit stops; upscale music venues; trendy bars and lounges; a diverse dining scene, which includes minimalist coffee shops; and community initiatives. International college students, young families taking advantage of excellent local public schools, elderly residents born and raised in Park Slope, and wealthy professionals searching for luxurious mansions and progressive attitudes all live alongside one another in Park Slope's brownstones (Manzo, 2012).

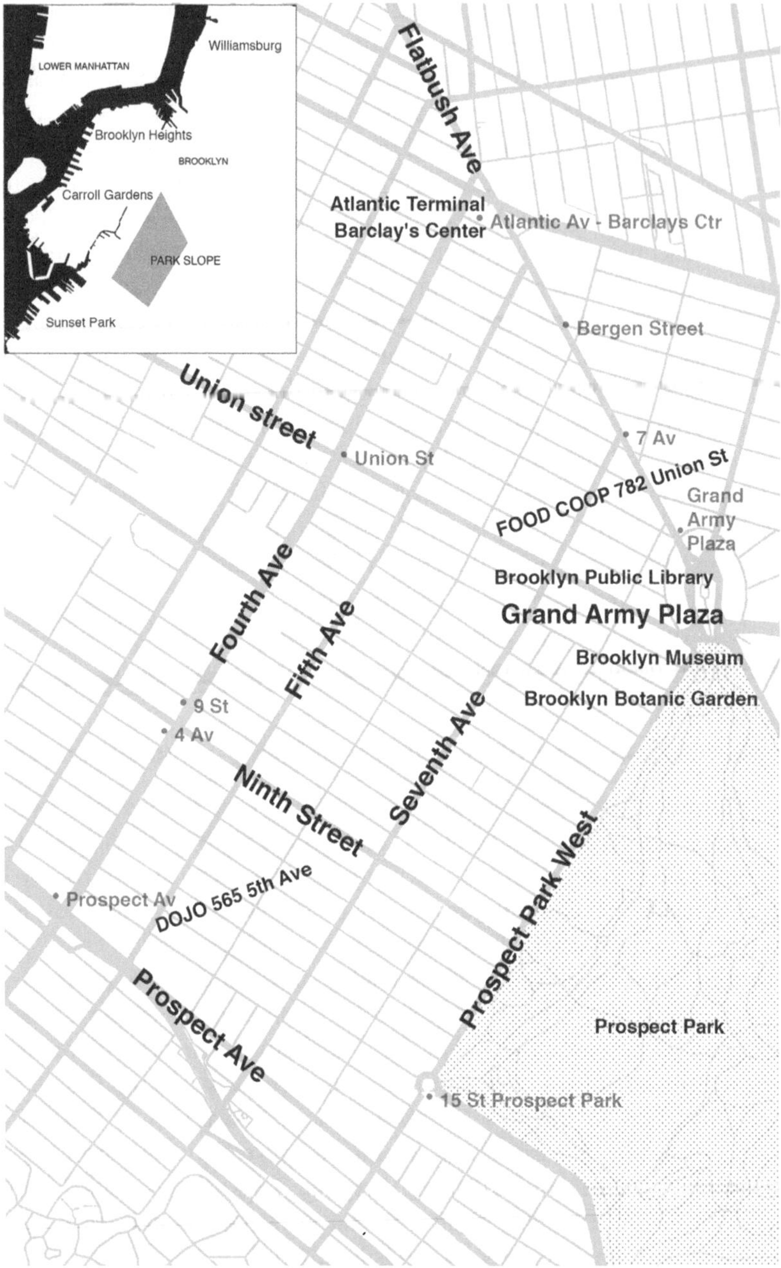

Figure 1. Brooklyn's park slope and its adjoining neighborhoods (Credit: Scott Lizama).

Park Slope has evolved from being predominantly White European (Irish, Polish, and Italian), White Hispanic, and Non-Hispanic Black to a White neighborhood. This transition occurred through an influx of White gentrifiers who do not identify with longstanding ethnic minorities (see Manzo, 2014). However, while the racial make-up of Park Slope has not changed much, neighborhood residents have become increasingly wealthier, better educated, and more heavily involved in professional occupations and in the financial services industry. The most dramatic change in Park Slope has been the increase in incomes, with the most rapid increase taking place during the 1980s and after the 1990s. Over the past 40 years, the median household income in Park Slope has increased by 1,162.8 percent – double the citywide rate.[5] These changes point to what Lees (2003) calls "super-gentrification."

In the following section, I introduce the two ethnographic settings where the study was conducted. These represent only part of a larger project that addressed the influences of displacement, housing-abandonment, and resettlement in the forty years that Park Slope was undergoing super-gentrification. The research consists of two and half years of ethnographic fieldwork mostly conducted in 2011 and 2012, with shorter follow-up studies in 2013, 2015 and 2017.

Tune your body up: the ethnographer as co-performer

The aim of the following section is aid the learning process by describing: 1) the choice of the two ethnographic settings; and 2) the strategies employed to gain access to them and the author's position as co-performer. This section concludes with a discussion of ethnographic data that provide evidence of performance as a paradigm-driven methodological tool that facilitates the study of embodied elements of elite reproduction.

Setting the stage: introducing the food co-op and the karate Dojo

One possibility for mitigating uncertainties of urban ethnography is to use fixed, real places with definitive boundaries; this tactic was used to select the case studies. Two distinct kinds of semi-private places were chosen along the lines of two criteria: both had to attract the economic elites of Park Slope as patrons, but they could not be the ruling actors. In such a way, the elite would have to interact with their lower class counterparts "equally". I found that the martial arts studio Japan Karate Association of Brooklyn and the food cooperative market Park Slope Food Co-op (PSFC) met the aforementioned criteria. Both were founded in the early 1970s and were built upon the values of democracy and member equality. Since then, they have welcomed members of multiplicity of genders, races, social classes, political affiliations, religious beliefs, and sexual orientations. For the purposes of this research, the assumption was made that one could find the entire spectrum of diversity within both spaces. Both had the potential to mirror the socio-demographic changes that occurred in the previous 40 years. As David, one of the elite members of the dojo pointed out,

> [k]arate, having its own set of rules, it is kind of a very neutral space. For instance, I know next to nothing about the lives of people of the dojo outside the dojo, and they're people who I probably wouldn't associate with outside that context. I might, but I don't know that I would. It's kind of cool because you'll have this group of people who are quite dissimilar from each other: very different backgrounds, very different ages, and very different professions. But they are all sort of brought together by this activity that they're doing together (...) you form some degree of a social connection with these people, even though there aren't other circumstances where that would happen. I mean I don't know when I would ever have a chance to interact with that kind of person socially. (« After training conversation », March.6.2015)

I also found them appropriate ethnographic settings for three important social reasons. First, the elite use the arts, sports, and "foodie-ism" as symbols of class prestige. Second, it is not particularly difficult for one to join one of these organizations. Membership requires the acceptance of their internal rules and payment of a membership fee (excluding the external issue of gaining access to elites' social network and sphere). Finally, I could enrich my research by joining an organization I had previously been involved with, as well as one that was completely new to me. Karate dojo, was already "familiar territory" (Hoggart, Lees, & Davies, 2002, p. 267) for me. Having been a karate practitioner for over twenty years and having had the chance to train already in that specific dojo,[6] I was confident in my ability to develop the network of research contacts through these practitioners and gatekeepers. On the other hand, knowing very little about the food co-op gave me the opportunity to understand the interactions and behavior from another perspective (van Maanen, 1988). The martial arts studio had a thirty adult member capacity, and the oldest American food cooperative had 16,000 members. Thus, I could observe how intimate relationships affect human behavior and interaction differently from vaster membership and diversity.

At the karate dojo, located on the southern end of Park Slope's Fifth Avenue, the master's teachings of *Shotokan* style are profound and exhaustive, with training that focuses on basics, form, and sparring. Karate practitioners must learn to work together with respect and courtesy, and they must always practice karate safely but effectively. "Throughout their study of karate, students are constantly forced to reflect upon themselves, their attitude and their behavior" (Umezawa, 1998, p. 10). The dojo is composed of a community of practitioners from a variety of class backgrounds and ethnicities. At the dojo, one's financial standing or class/educational background does not matter; one achieves stature by merit. The ritual of the gi, or uniform, is an equalizer. The only difference in uniform is the belt, which distinguishes one's rank. A dojo is not a temple, as karate is not a religion; however, there are some simple rules of respect. One should bow upon entering and exiting the dojo as well as help maintain and clean the space. Every member, including the master, must respect the five precepts of karate, the *dojo kun*.[7] These are as follows: 1) strive for perfection of character; 2) be faithful and protect truth; 3) endeavor; 4) respect others and the rules of etiquette; and 5) refrain from violent behavior.

The PSFC is the oldest and largest active American members-only food cooperative. It is located on Union Street. Its mission is, "Good Food, at Low Prices for Working Members, through Cooperation," and Park Slope residents (and local media outlets) describe it as a space "where elite lefties meet to share organic recipes as they gossip about social justice and the best private schools" (Peyser, 2011). Since 1973, it has been providing Brooklyn and beyond with quality food and products, while serving as a community center and meeting place for its member-owners: people who believe in the value, rewards, and responsibility of collective labor, action, and ownership. Similarly to the dojo, the democratically-controlled PSFC is founded upon five values: 1) equity and equality; 2) self-help; 3) self-responsibility; 4) democracy; and 5) solidarity.

Despite gentrification which has altered the landscape of different areas within Park Slope since the late 1970s (Manzo, 2014), both the karate dojo and the PSFC have maintained their lower-class members. Even former residents that were priced out of the neighborhood's rental market did not abandon training with their karate master or purchasing affordable, organic food. I observed members of both institutions traveling from all New York City boroughs and even Jersey City. These equality politics catered to lower class members through the provision of affordable or reduced membership fees and extended payment plans.

Gaining access to the field

The most important step of the research process was gaining access to the two settings in ways that were appropriate, ethical, and effective (Madison, 2005). To be successful there are no particular secrets beyond authentic, profound commitment, open-mindedness, and a reliable, sincere interest, which informants always seem to perceive in the researcher. Indeed, co-performance ethnography requires the researcher to also enter into a specific fieldwork situation with a "tuned-up body," as suggested by Goffman, to have "the right to be close to them" (1989, pp. 125–126) and access embodied knowledge.

At the karate dojo, I performed an adult, third *dan*[8] black belt, beginning in January 2011. As a skilled karate practitioner, the Brooklyn dojo allowed me to outline a preliminary framework of community understanding. However, being a female black belt in a place that is dominated by male practitioners, I had to show them that I was just as "tough." It took me two months of very hard training to earn "a position of respect" within the dojo. My new reputation, which I could have only obtained through my personal karate skills, had an inestimable value during the course of the data collection phase of my research.

I joined the PSFC in May 2011. I worked as a volunteer member in many different positions, including shopping assistant, cashier, co-counter attendant, and maintenance worker. I also sliced, prepared, and packaged cheeses, olives, spices, and dried fruits to be sold in the market.

Something that was unanticipated when I designed my research was the opportunity to be perceived very differently in the two settings. Due to my level of experience with karate I soon became part of the dojo advanced practitioners group, mostly composed of, "long-term working class residents." On the contrary, being a well-educated, White European woman in her late 30s with a different accent, mannerisms, and cultural background from the long-term Brooklynites, people at the coop thought of me as one of the, "upper class volunteer members." Following Katz's solution, I explored myself in the "spaces of betweenness" (1992, p. 506). I found explicit connections between the constant reflexive reposition and political practice of myself and those with whom I co-performed.

All research informants with whom I worked directly were aware of the purpose of my study and consensually shared their feelings, perceptions, and data about their socio-demographic background and lifestyles. The negotiation of trust revolved around the ethic: do not harm research participants and protect their anonymity. Nevertheless, to mitigate intrusiveness, I decided to record my observation in the research diary in privacy (Hoggart et al., 2002).

Co-performing off-stage elite practices

This section delves into the discussion of elitist signs of superiority, as analyzed in the two ethnographic settings. There is no method to reduce these signs to one single rationale because they represent a more complex frame. Nevertheless, a typology has been outlined to guide the reader through the categorization of performances on the path to creating symbolic boundaries (see Figure 2).

In the case of the karate dojo and the food co-op in Park Slope, elites have to "face" two distinct yet class-diverse group settings with similar norms of conduct based on the principles of democracy and equality. A major contradiction is that the aspiration toward such progressive moral values lead instead to hegemonic stylized acts of "exclusive behaviors" (Lamont, 1992, p. 10). As mentioned in the introduction, self-awareness is taken "off" in certain settings, and elites deviate from group norms to boost their self-esteem and reduce uncertainty (Abrams & Hogg, 2001; Hogg, 2001; Stets & Burke, 2000). This study showed that structural status-based differentiations were accentuated in places where elites exercise coercive power to remedy a perceived lack of power. Elitist performances resulted in either successful distinction and domination or anti-social practices and excessive frustration. The latter was the case when elites failed to exert power. The creation of symbolic boundaries is related to the definition of context. The elites attempted to define the context more broadly in order to incorporate that social sphere in which they are privileged. Non-elites attempted to define the context more narrowly to eliminate the disadvantages they face outside this social space.

Exertion of distinction and domination is evident in both settings, despite the fact that elites "take off" most of their external signs of wealth by changing into

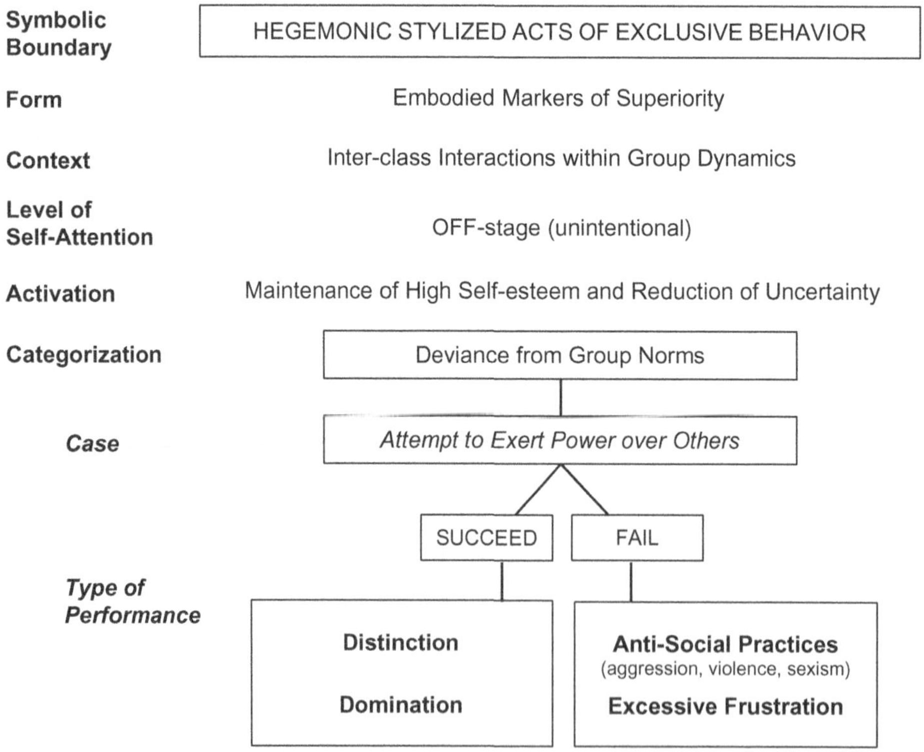

Figure 2. Typology of elites' performance of superiority in the production of symbolic boundaries.

uniforms. Situations in which elites can be the dominant actor are multiple. They carry out claims of prestige whenever possible. At the karate dojo, I observed this behavior between practitioners of different levels. Martial arts practice involves the acceptance of a hierarchical ladder of respect. The more experienced black belts are the most respected. The colored belts, which range from brown to white,[9] follow in order. This system allows beginners to display deference while learning the more advanced karate techniques from those higher up. While in principle all members agree on this practice, I observed that elites are less willing to take lessons from advanced practitioners of lower socio-economic status, who have lived in Park Slope since it was a "bad neighborhood." Many of these practitioners are construction workers, drivers, or nannies. Due to the evolutionary process of gentrification, it was more likely to find advanced black belts that were of lower socioeconomic status because they were long-term Park Slope residents and members of the dojo. On the contrary, the beginners mostly belonged to the "upper-class newcomers" social cohort. Even *Sensei*, the Japanese dojo Master, noticed that as the neighborhood population was changing in southern Park Slope, his karate students were changing too.

> Our dojo people… more intelligent people in the dojo, some PhD, or some managers, maybe 25 years, 30 years before, dojo people more criminals, drug dealers. Nowadays more educated people stay here. (Sensei, 68, karate Master)

Practicing karate in a place mostly populated by lower class advanced black belts endangered the elites' egos because they were not in the position of power. By co-performing karate across different courses offered in the dojo, I found a "class" boundary at play, both in terms of karate class and social class. Elites who were brown or first grade black belts strategically preferred to choose lower level courses in the *intermediate* class to avoid training with lower class black belts in the *advanced* one. Elites justified this behavior by stating they felt unprepared for the higher intensity of the advanced class, especially the sparring sessions.[10] On the contrary, I observed that the intermediate class allowed those elites to be the "superior" practitioners. Because very few advanced practitioners attended the intermediate classes, elites were usually the "head students." As a result, they were in charge of carrying out the most important class ritual: the bow or, in Japanese, *rei*. In order to perform it, students are asked to line up in a grid pattern: the highest ranks to the left, the next behind them, and so on down the line. Brown and first grade black belts led the bow line in intermediate classes, and they steered the meditation (*mokusoh*) and the initial recitation of the five principles of karate. They also usually led the warm-up and the final stretch sessions and, more importantly, they acted as a point of reference for the less-experienced students during class. However, their lower class counterparts perceived elites' situated self-meanings as diverging from the prototypical norms of the dojo, and they interpreted such distancing behavior as a sign of snobbery. As noted by Malloy and Frankie,[11] one day after class:

Malloy: The austerities are the training. It's the breaking down of the body so the mind will open. That is what we do in class. Practice. That's the environment that we have. Nobody really gets involved. Like people who… "oh don't touch my face or fuck you!" Who do you think you are? All of that shit goes out the window, you know what I mean.

Frankie: People get treated the same way, all that nonsense goes out the window, about your class and who you are and that's what make the Dojo. (« After training conversation », May.16.2012)

Elite practitioners often misunderstood the "teaching intentions" of the lower-class advanced black belts, interpreting them as signs of bullying or reverse discrimination. Many times, the elite confessed to me in the changing room or during private conversations that they were distressed by the actions of advanced belts. They thought these practitioners were "rude" to them and were "taking advantage of their low level of technique" in order to "put them down" or "beat them." Damian expressed this unease at the end of the training. He pointed out how elites behave normatively in the group and how lower-class members counteract the disturbance.

> Frankie was sparring today with Damian, who kicked him with a *mae geri* (front kick). As Damian threw a *mae geri* and went to step down, Frankie stepped back and accepted the kick. He allowed him to hit him, but his range was perfect because it didn't hit him 100%. He hit him about like maybe 35%. And then when he went to step down, Frankie went and swept him. And Damian fell down, seemingly embarrassed, like he didn't get "the lesson"! (« Field note », May.16.2012)

At the PSFC, I was confronted firsthand with the symbolic boundaries at play between the upper class volunteer members and the lower class workers. On the first day, I worked in the co-op's "receiving" area, where bad vegetables must be taken off the shelf and restocked. After a while, I went back to the receiving area to divide the produce into different bins. I was a bit irritated at that point, since my hands were dirty from rotten vegetables, and I was cold because I had been sorting them in the fridge for about an hour. At the same moment, another member noticed that there was a mess surrounding recycle bins; some of the salvageable vegetables had ended up in the compost trash instead of the soup kitchen one, or vice versa,

> Argh, and now I'm here with my stuff to put into these bins... what do I have to do? And while trying to fix this situation, here it is, the voice of the Coop coordinator, an African American woman, screaming at me. Oh no... Maybe she was already talking to me, and I didn't realize that because I was bent over the garbage, and probably this infuriated her. Still bent over, with the bad tomatoes in my hands, from the bottom up, I turn my face toward her voice, and she was there, above my head, yelling madly at me: "You see this mess? Now clean it all up!" (« Field note », June.21.2011)

I could not even say anything at that point and remained bent over. I lowered my gaze once again, and then I realized that I had made a big mistake. I probably did not reply as she expected or wanted because she continued to scream the same things. I thought:

> I don't believe this... I'm fixing this mess, what the hell does she want from me? She knows that I'm new, and I didn't make this mess! [I want to cry...] but suddenly I understood: I disrespected her, she wanted to be revered!

And once again, I turned my face upward to look at her, finally saying what she needed to hear:

> "Alright!"

I was really right, since she calmed down and left me alone, in front of the garbage. That day, there was something deeper to be analyzed than simply an asymmetry in power. I never knew the term "alright" could be used as a sign of surrender and acceptance. In addition, having most likely been deemed an upper-class co-op member, I had the chance to experience firsthand how staff of Caribbean, African American, or Puerto Rican background treated White gentrifiers differently than their long-term-Park Slope-resident counterparts. For example, at the end of my shift that day, I noticed that there were cotton gloves and aprons for members to use while doing "dirty work" in a small box in the receiving area. Even though New Yorkers usually follow the "every man for himself"[12] work ethic, I found it odd that nobody told first-time volunteers like me or the other young White woman, who was wearing a large flowery skirt and a pair of red ballerinas, how to avoid getting dirty. As I discussed in the case of the dojo, the co-op worker was unfriendly to me, underlining the "fine line" among the competing selves we are constantly performing. Our performance is dependent on the intersection of social class, ethnicity, gender, and also membership in local institutions like a karate dojo or a food co-op.

In addition, by holding different volunteering positions at the PSFC, I was able to observe many forms of social stratification: the higher one's social class (and level of education), the more likely he or she was to volunteer as cashier or administrator, as opposed to working in the food packaging or maintenance areas. Despite the values of equality that permeate the PSFC, elites tend to reproduce their class status by applying for positions, like the group coordinator (or "squad leader"), that involve more responsibility. Paradoxically, the democratic PSFC's microcosm seems to provide "the stage" for elites to re-enact class divide and exercise power, as confirmed by Edna, a long-term coop member:

> I was working in the office, and there was this woman coordinating, who is the most miserable, nasty person. I am a trained person, I am a therapist, I've been in group therapy, I've lived in a commune, I have faced all kinds of difficult people in my life... but, by the end of that two and a half hours I was ready to quit the Co-op! I was practically crying! She was just a bitch! She had a tone in her voice, I can't remember... but, I just remember the power! (Edna, 64, Psychotherapist).

In a private interview, one of the co-op's founders, Justin, did not deny the existence of "different styles" of volunteering. He thereby silently reaffirmed, that elites at the PSFC perform various facets of social distinction and how exclusionary behavior functions to reinforce and clarify groups' norms.

> Some people are abusing the squad leader role, like they're on a power trip sometimes. "This is my three hours every 28 days I get to be this, I am the leader" I know, and the Co-op is tolerant of many different styles of that. We don't micromanage that, I understand there are different styles. (Justin, 62, PSFC Manager)

With regard to the second situation, when the elites cannot be superior to ordinary people, aggression, violence, sexism, or excessive frustration may emerge. At the karate dojo, despite members usually getting along and enjoying after-training talks and company, elites displayed difficulty in accepting failure in sparring sessions. In the advanced class, when the social hierarchy is somehow reversed, elites suffer from the swift indignity of being treated as weak. They prefer to perform the kata (individual form) instead of working in pair for the kumite (sparring). If in the first they can manage a theatrical "impression of trained excellence" of the karate gestures, in the latter they have to confront the other fellows "face to face." It is the very physicality of kumite that contravenes the habit of good manners through which elites are socialized, as I discussed with some advanced fellows one time after class:

> In the past, it was more physical. I could tell you that for sure. It was more physical in the sense that you know, people who are used to sparring with each other of course would... don't mind hitting each other. They know that they're testing their limits. But because of people that are not physically, you know, aggressive or physically strong, the hitting seems not allowed now but they're trying to help you be better, they're not trying to hurt you. But they will hit you just to you, to try and that's right. (« After training conversation » with Jack, February.17.2012)

Frequently, a fear of failure has been observed as disruptive to their embodied sense of moral superiority. This leads to a loss of self-confidence, acted out through loss of emotional control, loss of relaxed attitude, and anti-social behavior. I saw elites training beside me become more and more red-faced and angry because they could not block the blows or attack their opponent. I saw their faces becoming tauter as the pressure of the blows made them lose their posture and sometimes their balance. I observed them crying under the grin of their opponents, but the observation that struck me the most was the constant attempt of the elites to "win at all costs." They always tried to overcome fighting exercise rules with the purpose of beating their opponents, whom usually – being advanced fellows – could anticipate those intentions and in turn fight the "astute elites" even harder. Elites deviate from the group norms of a specific context by asserting their class attitudes, thereby threatening the prototypical integrity of the group. Under these circumstances, an intragroup dynamic between fellow lower class in-groupers emerges as they attempt to reassert their position of influence. I also experienced first-hand attempts from elite men practitioners to enact a disproportionate use of physical strength on the black belt women, in order to be "at least better than them," a strategy that was not always useful. Overexerting oneself makes the athlete lose speed and balance for smooth transitions.

Another exemplary case at the PSFC involved a Greek businessman, Timo, who actually looks like he could be Latino. Together with Allison, a famous novel writer who belongs to the neighborhood's celebrities group, we were all taking a small break from our volunteer shift when he engaged in a discussion about a perceived abusive behavior from a co-op worker in the receiving area:

> So I said to the supervisor – and he was talking to three girls – "excuse me, can you show me how to use this machine?" And he turned around and he said "well you better learn." And you know I was surprised because it's the Co-op, and I didn't expect to be treated disrespectfully, and I said, well that was the point of me asking you how to use it, is that I need to learn how to use it. And then he said something to me in Spanish. And I said, I'm sorry I don't speak Spanish. And again he said "well you better learn, you're in Brooklyn!" and I was like, I know where I am and I'm Greek, I don't have to learn Spanish for you or anybody else (...) You expect that kind of behavior in the street in the Brooklyn or somewhere else but not at the Co-op! (Timo, 42, Self-employed).

According to Hogg, structural social class distinction affects intragroup relationships (and the making of boundaries) by "creating subgroups that may have competitive relations" (2001, p. 71). Aside from the fact that the manager was being overtly flirtatious with some of the women, and rude to Timo, the blow to his self-esteem and lack of verification motived Timo's performance. His gestures, tone, and his reddened face fumed with excessive frustration while demanding for the "usual" respect (or flattery) received at the coop. He also expressed a visible sense of repugnance for having being confused for a member of Brooklyn's Latino population.

As was pointed out in the theoretical introduction of this paper, ethnography as a "performative space for the full-throated acting out of everything that must be chocked back in public" (Conquergood, 1992, p. 91) is a key for the interpretation of those daily "zones of contact" between dominant elites and lower class citizens. During those "off-stage" moments, elites perform their contradictory intentions, experience misunderstandings and confusion brought by urban diversity, and, as a result, they express their sense of superiority.

Discussion and conclusions

In the super-gentrified neighborhood of Brooklyn's Park Slope, economically elite residents became members of democratic associations, like karate dojos or food co-ops, to be socially accepted by and mingle with the local communities. These newer residents proudly exhibited appropriate progressive moral values. However, in the act of undressing themselves from exterior signs of class status, deeply-rooted interior markers of superiority were revealed in the performative off-stage practices played out in those settings. At the micro-level, an analysis of these embodied attitudes provided evidence of an understanding of the reflexive self. The research confirmed how in-group performances are strongly influenced by class diversity. Elites addressed the distress faced when the self is constrained to different identities (class, group, role) by identifying themselves on the basis of their social class. On the other hand, lower-class members relied on meaning given by specific role-based identities. Through the production of symbolic boundaries, deviance and minority subgroups alike engaged in differential performances, relationships, and interconnections of roles (Stets & Burke, 2000). Ultimately, elites elicited internalized repetition of hegemonic stylized acts of exclusive behavior, which served as a basis for inequality and conflict. Sometimes they felt *good* and *confident* about performing distinction and dominations. At other times, their roles *constrained* them, creating conditions under which elites engaged in anti-social practices and demonstrated excessive frustration. These actions highlighted the complexity of elites' contrasting need to simultaneously assimilate and distinguish themselves in urban communities. It is also worth noting that the karate dojo and the PSFC represented significant "counter-places" in which lower class members could construct counterhegemonic discourses and practices of resistance.

Understanding how elites draw symbolic boundaries is also central to the very political view of performance as a space for agency and resistance to power, challenging critical geographers to engage in this research method to reinforce our knowledge of how issues of distinction, segregation, and hegemony work spatially. Especially at the community level, these micro-ethnographies narrate the process of negotiation that enables living together, debunking the existence of emotional and moral distress among people of different social groups experiencing super-gentrification. Social mixing in urban neighborhoods has brought about the constant performance of competing selves participating in the production of boundaries, depending on "both as we label others and we participate in communities whose shared beliefs make a specific definition of reality intersubjectively true" (Lamont, 1992, p. 10). According to Marcus, the study of elites does not confirm "who the good and bad guys are" (1998, pp. 27–28) between the rulers and ruled and the leaders and the led. Rather, it reveals the complexity and ambiguity of their relations of domination and moral position in everyday practices in local communities of difference.

Performance ethnography has proven to be an effective method to study intersubjectivity, allowing the hidden, deeply-rooted signs of elite distinction and the tacit symbolic boundaries to transpire. This approach clearly highlights how relations of power shape elites' lives and are produced by them (Manzo, 2015), calling for a radical rethinking of the hermeneutical agency of the body. Foucault's well-known argument (1995) is that embodied practices open up an interiority and therefore constitute knowledge. This notion resembles Gramsci's idea of the politically-engaged knowledge production of the intellectuals (1971), which must be linked to the feelings and passions of the people to make counter-hegemonic representations emerge. Therefore, the staging of elites' identities, behavior, and emotions raises important connections between body and capital accumulation. Such is especially relevant in reference to those embodied markers by which elites demonstrate their affluent social status rooted in the regime of the capital that created it (Longhurst, 2000).

Notes

1. Interactionists have been concerned with the ways in which subjective meaning emerges from encounters in situations that are always shifting and ambiguous (Blumer, 1969), socially constructed (Berger & Luckmann, 1966), and where personal identity is formed (Mead, 1934).
2. By practice Bourdieu refers to social patterns of behavior and habitus, "durable, transposable dispositions" (1977, p. 72) by which individuals contribute to the reproduction of structure of domination.
3. These behaviors involve matters of etiquette, dress, deportment, gesture, intonation, dialect, vocabulary, small bodily movements and automatically expressed evaluations concerning both the substance and the details of life (Goffman, 1951, p. 300).
4. Prospect Park contains recreational areas, a zoo, a stage, ponds, a lagoon, picnic grounds, and different sport pitches.
5. Source: Author's elaboration on U.S. Census Bureau data for Park Slope and New York City, 1970–2010.
6. In 2009 I decided to spend a month on vacation in New York City, and I found a room for rent in what, at the time, I did not know would become the object of my research in 2011: The neighborhood of Brooklyn's Park Slope.
7. Developed by the founder Master Gichin Funakoshi.
8. The dan is the ranking system used in karate and other martial arts to indicate the level of one's ability.
9. Single character variations, in terms of disrespectful attitudes, are commonly managed by the Master, who in case of misbehavior can decide to withdraw the dojo membership.

10. As I will explain in the section devoted to the anti-social behavior, this attitude is not very "martial," as usually it is considered an honor and a great opportunity to train with experienced practitioners.
11. All informant names have been fictionalized in order to protect anonymity.
12. As a way to say that typically co-workers do not help each other out.

Acknowledgments

This research would not have been possible without the generosity of the Food Coop members and the Karate Dojo fellows who took part in this ethnography in the neighborhood of Brooklyn's Park Slope. I also am particularly grateful to Susan Moore for her support and direction as editor, and the three anonymous reviewers whose input has significantly improved this article. Further thanks to David Bassens, Enzo Colombo, Mary Gilmartin, and Bas van Heur for their comments on various iterations of this work.

Disclosure statement

No potential conflict of interest was reported by the author.

Funding

Research and writing for this article was supported by the Irish Research Council post-doctoral fellowship scheme at Maynooth University, Ireland (grant GOIPD/2015/518) and by a Milan City Council doctoral grant-in-aid awarded at The University of Trento, Italy.

ORCID

Lidia Katia Consiglia Manzo http://orcid.org/0000-0001-8530-6719

References

Aalbers, Manuel B. (2008). The financialization of home and the mortgage market crisis. *Competition and Change*, *12*(2), 148–166.
Aalbers, Manuel B. (2016). *The financialization of housing. A political economy approach*. London: Routledge.
Abrams, Dominic. (1996). Social identity, self as structure, and self as process. In William Peter Robinson (Ed.), *Social groups and identities: Developing the legacy of Henri Tajfel* (pp. 143–168). Oxford: Butterworth-Heinemann.
Abrams, Dominic, & Hogg, Michael A. (2001). Collective identity: Group membership and self-conception. In Michael A. Hogg & Scott Tindale (Eds.), *Blackwell handbook of social psychology: Group processes* (pp. 425–460). Malden, MA: Blackwell.
Andreotti, Alberta, Le Galès, Patrick, & Moreno-Fuentes, Francisco Javier. (2015). *Globalised minds, roots in the City: Urban upper-middle classes in Europe*. Hoboken, NJ: Wiley-Blackwell.
Atkinson, Anthony B., & Piketty, Thomas. (2010). *Top incomes: A global perspective*. Oxford: Oxford University Press.
Auslander, Philip. (2008). *Theory for performance studies*. New York: Routledge.
Beaverstock, Jonathan V., Hubbard, Phil, & Short, John Rennie. (2004). Getting away with it? Exposing the geographies of the super-rich. *Geoforum*, *35*(4), 401–407.
Bennett, Tony, Savage, Mike, Silva, Elisabeth, Warde, Alan, Gayo-Cal, Modesto, & Wright, David. (2009). *Culture, class, distinction*. New York: Routledge.

Berger, Peter L., & Luckmann, Thomas. (1966). *The social construction of reality*. London: Penguin.
Blumer, Herbert. (1969). *Symbolic interactionism: Perspective and method*. Englewood Cliffs, NJ: Prentice-Hall.
Borsdorf, Axel, & Hidalgo, Rodrigo. (2008). New dimensions of social exclusion in Latin America: From gated communities to gated cities, the case of Santiago de Chile. *Land Use Policy*, *25*(2), 153–160.
Bourdieu, Pierre. (1977). *Outline of a theory of practice*. New York: Cambridge University Press.
Bourdieu, Pierre. (1984). *Distinction: A social critique of the judgement of taste*. Cambridge, MA: Harvard University Press.
Bridge, Gary, Butler, Tim, & Lees, Loretta. (2012). *Mixed communities: Gentrification by stealth?* London: PolityPress.
Butler, Judith. (1988). Performative acts and gender constitution: An essay in phenomenology and feminist theory. *Theatre Journal*, *40*(4), 519–531.
Butler, Tim, & Lees, Loretta. (2006). Super-gentrification in Barnsbury, London: Globalization and gentrifying global elites at the neighbourhood level. *Transactions of the Institute of British Geographers*, *31*(4), 467–487.
Clifford, James. (1988). *The predicament of culture*. Cambridge, MA: Harvard University Press.
Clifford, Nicholas, Cope, Meghan, Gillespie, Thomas, & French, Shaun (Eds.). (2016). *Key methods in geography*. London: Sage.
Conquergood, Dwight. (1989). Poetics, play, process, and power: The performative turn in anthropology. *Text and Performance Quarterly*, *9*(1), 82–88.
Conquergood, Dwight. (1991). Rethinking ethnography: Towards a critical cultural politics. *Communication Monographs*, *58*(2), 179–194.
Conquergood, Dwight. (1992). Ethnography, rhetoric, and performance. *Quarterly Journal of Speech*, *78*(1), 80–97.
Cousin, Bruno, & Chauvin, Sébastien. (2013). Islanders, immigrants and millionaires: The dynamics of upper-class segregation in St Barts, French West Indies. In Iain Hay (Ed.), *Geographies of the super-rich*. Cheltenham: Edward Elgar.
Cousin, Bruno, & Chauvin, Sébastien. (2014). Globalizing forms of elite sociability: Varieties of cosmopolitanism in Paris social clubs. *Ethnic and Racial Studies*, *37*(12), 2209–2225.
Daloz, Jean-Pascal. (2010). *The sociology of elite distinction: From theoretical to comparative perspectives*. Basingstoke: Palgrave Macmillan.
Denzin, Norman K. (2014). Cultural struggles: Performance, ethnography, praxis by Dwight Conquergood, auth., and E. Patrick Johnson, ed. *American Anthropologist*, *116*(3), 676–677.
Durkheim, Émile. (1965). *The elementary forms of religious life [1912]*. (Joseph Ward Swain, Trans.). New York: The Free Press.
Fernandez, Rodrigo, Hofman, Annelore, & Aalbers, Manuel B. (2016). London and New York as a safe deposit box for the transnational wealth elite. *Environment and Planning A*, *48*(12), 2443–2461.
Fields, Desiree. (2015). Contesting the financialization of urban space: Community organizations and the struggle to preserve affordable rental housing in New York City. *Journal of Urban Affairs*, *37*(2), 144–165.
Foucault, Michel. (1995). *Discipline and punish: The birth of the prison* (2nd ed.). New York: Vintage Books.
Goffman, Erving. (1951). Symbols of class status. *The British Journal of Sociology*, *2*(4), 294–304.
Goffman, Erving. (1956). *The presentation of self in everyday life*. Edinburgh: University of Edinburgh.
Goffman, Erving. (1983). The interaction order. *American Sociological Review*, *48*(1), 1–17.
Goffman, Erving. (1989). on Fieldwork. *Journal of Contemporary Ethnography*, *18*(2), 123–132.
Gramsci, Antonio (1971). *Prison notebooks selections*. (Q. Hoare & G. Smith, Eds.). New York: International.
Gregson, Nicky, & Rose, Gillian. (2000). Taking Butler elsewhere: Performativities, spatialities and subjectivities. *Environment and Planning D: Society and Space*, *18*(4), 433–452.

Hay, Iain. (2013). *Geographies of the super-rich.* Cheltenham : Edward Elgar.
Hogg, Michael A. (2001). Social categorization, depersonalization, and group behavior. In Michael A. Hogg & Scott Tindale (Eds.), *Blackwell handbook of social psychology: Group processes* (pp. 56–85). Malden, MA: Blackwell.
Hoggart, Keith, Lees, Loretta, & Davies, Anna. (2002). *Researching human geography.* London: Arnold.
Hutter, Mark. (2012). *Experiencing cities.* Boston, MA: Pearson.
Jackson, Kenneth T. (2004). *The neighborhoods of Brooklyn* (Vol. 1998). New Haven: Yale University Press.
Jackson, Michael. (1989). *Paths toward a clearing: Radical empiricism and ethnographic inquiry. Journal of Chemical Information and Modeling.* Bloomington: Indiana University Press.
Katz, Cindy. (1992). All the world is staged: Intellectuals and the projects of ethnography. *Environment & Planning D: Society & Space, 10*(5), 495–510.
Keister, Lisa A., & Lee, Hang Young. (2014). The one percent: Top incomes and wealth in sociological research. *Social Currents, 1*(1), 13–24.
Krippner, Greta R. (2005). The financialization of the American economy. *Socio-Economic Review, 3*(2), 173–208.
Lamont, Michèle. (1992). *Money, morals, and manners: The culture of the French and American upper-middle class.* Chicago: University of Chicago Press.
Lamont, Michèle, & Fournier, Marcel (Eds.). (1992). *Cultivating differences: Symbolic boundaries and the making of inequality.* Chicago: University of Chicago Press.
Lamont, Michèle, & Molnár, Virág. (2002). The study of boundaries in the social sciences. *Annual Review of Sociology, 28*(1), 167–195.
Lees, Loretta. (2000). A reappraisal of gentrification: Towards a "geography of gentrification". *Progress in Human Geography, 24*(3), 389–408.
Lees, Loretta. (2003). Super-gentrification: The case of Brooklyn Heights, New York City. *Urban Studies, 40*(12), 2487–2509.
Lees, Loretta. (2008). Gentrification and social mixing: Towards an inclusive urban renaissance? *Urban Studies, 45*(12), 2449–2470.
Lees, Loretta. (2013). The urban injustices of new labour's "new urban renewal": The case of the aylesbury estate in London. *Antipode, 46*(4), 921–947.
Lees, Loretta, & Ferreri, Mara. (2016). Resisting gentrification on its final frontiers: Learning from the Heygate Estate in London (1974–2013). *Cities, 57*, 14–24.
Lees, Loretta, Shin, Hyun Bang, & Lopez-Morales, Ernesto (Eds.). (2015). *Global gentrifications: Uneven development and displacement.* Bristol: Policy Press.
Lees, Loretta, Shin, Hyun Bang, & Lopez-Morales, Ernesto. (2016). *Planetary gentrification.* Cambridge: Polity Press.
Longhurst, Robyn. (2000). *Bodies: Exploring fluid boundaries.* London: Routledge.
Madison, D. Soyini. (2005). *Critical ethnography: Method, ethics, and performance.* Thousand Oaks, CA: SAGE.
Manzo, Lidia Katia Consiglia. (2012). On people in changing neighborhoods. Gentrification and social mix: Boundaries and resistance. A comparative ethnography of two historic neighborhoods in Milan (Italy) and Brooklyn (New York, USA). *Cidades, Comunidades E Territórios, June*(24), 1–29.
Manzo, Lidia Katia Consiglia (2014). *«Give me a break! I'm from Brooklyn, we're not fancy». Institutions, Housing and Lifestyles in Super-gentrification process. A Field and Historical research in Park Slope, New York City.* (Unpublished PhD thesis). The University of Trento, Faculty of Sociology.
Manzo, Lidia Katia Consiglia. (2015). And their struggle becomes visible. For a radical revaluation of Foucault's conception of resistance to power. In Lidia Katia Consiglia Manzo (Ed.), *Culture and visual forms of power. Experiencing contemporary spaces of resistance* (pp. 1–10). Champaign, IL: Common Ground Publishing.
Marcus, George E. (1998). *Ethnography through thick and thin. Katalog BPS* (Vol. XXXIII). Princeton: Princeton University Press.

Mead, George Herbert. (1934). *Mind, self & society from the standpoint of a social behaviorist.* Chicago: University of Chicago Press.
Myers, Norman, & Kent, Jennifer. (2004). *The new consumers: The influence of affluence on the environment.* Washington, DC: Island Press.
Osman, Suleiman. (2011). *The invention of brownstone Brooklyn.* New York: Oxford University Press.
Peyser, Andrea. (2011, August 25). Goy-cott in park slope. *The New York Post*, New York.
Pow, Choon Piew. (2011). Living it up: Super-rich enclave and transnational elite urbanism in Singapore. *Geoforum, 42*(3), 382–393.
Richards, Alison. (2001). Shaking the frame: Erving Goffman and performance studies. *Australasian Drama Studies, 39*, 58–75.
Rose, Damaris, Germain, Annick, Bacqué, Marie-Hélène, Bridge, Gary, Fijalkow, Yankel, & Slater, Tom. (2013). "Social Mix" and neighbourhood revitalization in a transatlantic perspective: Comparing local policy discourses and expectations in Paris (France), Bristol (UK) and Montréal (Canada). *International Journal of Urban and Regional Research, 37*(2), 430–450.
Sassen, Saskia. (2014). *Expulsions: Brutality and complexity in the global economy.* Cambridge, MA: The Belknap Press of Harvard University Press.
Simmel, Georg (1950). *The sociology of Georg Simmel [1903].* (Translated, edited, and with and introduction by Kurt H. Wolff). New York: The Free Press.
Slater, Tom (2003). Comparing gentrification in South Parkdale, Toronto and Lower Park Slope, New York City: A "North American" model of neighbourhood reinvestment? *CNR Paper 11.* Retrieved from http://www.urbancentre.utoronto.ca/pdfs/curp/CNR_Parkdale-NYC-Gentrifica.pdf
Slater, Tom. (2006). The eviction of critical perspectives from gentrification research. *International Journal of Urban and Regional Research, 30*(4), 737–757.
Slater, Tom. (2009). Missing Marcuse: On gentrification and displacement. *City, 13*(2), 292–311.
Smith, Neil. (1996). *The new urban frontier: Gentrification and the Revanchist City.* London: Routledge.
Smith, Neil, & DeFilippis, James. (1999). The reassertion of economics: 1990s gentrification in the lower east side. *International Journal of Urban and Regional Research, 23*(4), 638–653.
Stets, Jan E., & Burke, Peter J. (2000). Identity theory and social identity theory. *Social Psychology Quarterly, 63*(3), 224–237.
Story, Louise, & Saul, Stephanie. (2015, February 7). Stream of Foreign wealth flows to Elite New York real estate. *The New York Times.*
Thrift, Nigel. (2008). *Non-representational theory: Space, politics, affect. The dictionary of human geography.* New York: Routledge.
Turner, Victor. (1975). *Revelation and divination in Ndembu ritual. Man* (Vol. 12). Ithaca: Cornell University Press.
Turner, Victor. (1986). *TheAnthropology of performance.* New York: PAJ Publications.
Turner, Victor, & Bruner, Edward M. (Eds.). (1986). *The anthropology of experience.* Urbana: University of Illinois Press.
Umezawa, Rui. (1998). *The empty hand: A karate word book.* New York: Weatherhill.
van Maanen, John. (1988). *Tales of the field: On writing ethnography.* Chicago: University of Chicago Press.
Veblen, Thorsten. (1899). *The theory of the leisure class.* New York: Viking Press.
Zukin, Sharon. (1982). *Loft living. Culture and capital in urban change.* Baltimore: The Johns Hopkins University Press.
Zukin, Sharon. (2010). *Naked city. The death and life of authentic urban places.* New York: Oxford University Press.
Zukin, Sharon, Kasinitz, Philip, & Chen, Xiangming (Eds.). (2015). *Global cities, local streets: Everyday diversity from New York to Shanghai.* New York: Routledge.

OPEN ACCESS

Contingently elite: affective practices of diasporic urban nightlife consumption

Lauren B. Wagner

ABSTRACT

For diasporic communities fostered through international labor migration, visiting an ancestral homeland can be a transformative encounter. Crossing into "the homeland", descendants of migrants can reorient from a relatively underprivileged and socioeconomically immobilized minority, into a geopolitically mobile economic elite. For Moroccans from Europe, this transformation recurs every summer, as millions of diasporic visitors – including several family generations post-migration – travel "home" for summer vacation. These visits are as much an investment in belonging "at home" as they are a chance to consume leisure relatively inexpensively – to pursue, for a finite period in this familiar, familial place, affective practices of comparatively elite nighttime urban leisure available to them in Morocco more so than in Europe. As they move between and gather in consumption sites that are relatively inexpensive to them as diasporic visitors, they become a critical mass of nightlife consumers moving around cities in Morocco, becoming emergently and contingently elite.

Perception of perception of arrogance

> S: you know when I eh phh, it depends(.) sometimes when I go to other countries, then I miss Morocco, but when I go to Morocco I come back I said oh why did I go to Morocco, the people are ennerving me, *ils m'ennervent*
>
> [3 turns excised]
>
> LW: what is annoying you?
>
> S: well phhhh, what is annoying me, the fact that they think that we in Europe are millionaires here, and we don't have to work for our money and so(.) the fact that they think that we are arrogant, we are not arrogant.

In an interview that took place in Antwerp, Soumia (above) describes one circulating perception that influences her experience of Morocco as a "diasporic visitor" (DV), having been raised in Belgium within a Moroccan-origin family. She links her irritation – her annoyance, as I translated "ennerving" in situ – to a perception she attributes generically to local residents ("they") of thinking that DVs ("we in Europe") are arrogant. That arrogance is made relevant by (her assertion of) "their" misperceptions about eliteness: that she and other DVs visiting Morocco from Europe do not need to "work for [their] money"; that they occupy a position of leisurely wealth.

Her report of being perceived as arrogantly elite was echoed by other DVs – and has been echoed by other diasporically-configured visitors in other homelands (Coles & Timothy, 2004; Potter, Conway, & Phillips, 2005; Stephenson, 2002) – as an attitude they know circulates about them, but they are unable to effectively ameliorate. Even if their intention is not, as another participant described it, to "raise themselves a degree above others", they reported being aware of perceptions about them as arrogantly elite; perceptions they attributed to local residents. Effectively, their interactions outside their families (and sometimes within them) can become a vicious cycle of negative perceptions reproducing each other: DVs indignantly perceiving themselves as perceived as arrogant, and local residents apparently perceiving themselves as affronted by arrogant behavior. These negative perceptions rebound and accumulate into a affectively practiced dynamic of elite distinctiveness, in which these DVs, when feeling unjustly labeled as "millionaires," may choose – like Soumia threatens above – no longer to visit Morocco in order to avoid this negative experience.

Though Soumia seems to disavow participating in any "arrogance" herself, behavior that might be interpreted as arrogantly elite is implicitly part of how DVs are present in Morocco. As visitors who come only during holiday periods, often in search of experiential leisure as a key activity to enjoy their visit, they seem to suddenly explode into the urban nighttime economy as voracious and mobile leisure consumers. In comparison to some local residents, they seem to spend money on leisure without limits, at times unaware of implicit socially-organized standards, and occupying significant resources in public space as a cumulative critical mass. Though in their European places of residence they may encounter problems of access in participating in nighttime leisure in cities (Schwanen, van Aalst, Brands, & Timan, 2012), their status in Morocco as mobile, Muslim vacationers enables a freedom of movement and leisure consumption that they may not find elsewhere.

Through this situated contrast, this paper investigates the leisure consumption practices that shape how eliteness emerges for a cohort of post-migrant generation diasporic visitors, coming into contextual relevance through interactions with others in urban consumption spaces rather than absolutely measurable through metrics of income or wealth. The reported perception of their eliteness at "home" may contrast their situational non-eliteness in their countries of residence, where they are often contextualized alongside other stigmatized and ethnicized migrant minority communities, but it may also be accurate to how, in their leisure practices, they become present and perceivable as a mobile collective in Moroccan urban nightlife.

To that end, this paper begins by situating how this eliteness is configured by Moroccan histories, DV mobilities, and the production and consumption of urban nightlife in Morocco. Next, the discussion addresses the production of an elite distinction as an affective practice, and how it is empirically approached here using ethnographic methods. Finally, I present some fieldwork data, composed of fieldnotes paired with visuals and sketched maps, to illustrate how certain practices of urban leisure consumption, which were repeatedly observed across participants of many ages and in different cities throughout this project but are reported here through one exemplar "night out", might be read as affective practices materializing into perceptions of arrogant eliteness, cutting a pathway of diasporic leisure consumption through a Moroccan city.

Elite ecologies of summertime in morocco: night time experiential leisure across urban space

Eliteness as a practiced position incorporates the ability to define rules, create boundaries, and determine access to power (Myers-Scotton, 1990). Empirically, these practical attributes contour how others can situate themselves in relation to those who become "elite": while others may know they are nearby, or be able to trace evidence of their presence and influence, elites are rarely blatantly visible or directly accessible. Part of practicing eliteness may be in this invisibility, inasmuch as those who have no need to raise their status (financially or socially) also have no need to demonstrably show their status (Freudendal-Pedersen, 2014; Spence, 2016). Part of this practiced invisibility may be spatial, through the purposefully limited accessibility of enclosed spaces where elites are to be found – especially those spaces codified and practiced as protected, silent, and satisfying in order to attract elite consumers (Thurlow & Jaworski, 2012).

Yet, the "elites" under discussion here are not among those who might be generally considered, across diverse times and spaces, as defining rules, creating boundaries, and determining access to power. They are, instead, those who encounter frictions, disagreements and ambiguity in being considered as "elite" (Jansson, 2016). In line with how this this special issue suggests approaching urban eliteness through "ecologies" rather than classifications – bringing attention to how eliteness works to contour urban life as relational between many interacting parts rather than as a universal definition (Bassens and Van Heur, this volume) – the "eliteness" argued for here is dependent and contingent on a combination of factors occurring in certain limited timespaces. It is not likely to be extendable beyond that limited space and its set of contingencies. The collection of factors that make up this urban nightlife scene all contribute to how this collective emerges as "elite" in their contingent ability to test and break rules, enact boundaries, and exercise economic power – while, however, remaining starkly visible in the Moroccan urban nightlife environments where their "eliteness" is practiced.

Morocco presents perhaps a unique case in the production of urban space for consumption, to the extent that the "preservation" of each *medina*, or historical city, alongside the creation of *villes nouvelles* full of hotels and cafes, was explicitly part of the French colonial Protectorate project to engender and promote touristic consumption of the territory by a European public (Hillali, 2007; Minca & Wagner, 2016; Stafford, Bélanger, & Sarrasin, 1996). Modes of consumption encouraged through this dynamic, in parallel with other post-colonial sites for tourism (cf. Michael & Tucker, 2004), range from a discourse of authenticity linked to global heritage, to an invitation for embodied experience of "the local" adapted for the non-local consumer, to the promised pleasure found in natural landscapes, city centers and urban nightlife throughout the territory. In Morocco, this Protectorate-era legacy is perpetuated in the current King's development plans (Vision 2010, Vision 2020) predicated on tourist entries and nights of lodging as a key source of national income, and encouraging (with investment) the ongoing and increasing development of several Moroccan cities and landscapes as "unique" and "authentic" destinations for a wealthy foreign consumer

(Hillali, 2007). While many tourists from Europe and elsewhere consume Morocco as a site, many diasporic Moroccans partake of the touristic leisure sites that have proliferated in this development goal, oriented towards a "foreign" consumer but implicating their own consumption preferences during their summer holidays as well. The contours of possibilities for consuming urban leisure, specifically nighttime economies oriented towards a comparatively elite consumer, become essential to framing how eliteness emerges as a key distinctive element for Moroccan diasporic visitors.

Moroccan-origin communities across Europe, primarily composed of families who relocated during post-war labor migrations, reflect a recurrent model in migration studies of comparative economic success paired with problematized minority status. This wave of migrants were generally seeking economic solvency by finding work elsewhere, and for the most part achieved that in Europe; however, many communities suffered stigmatization, and many families are among the less wealthy in their European homelands (Chattou, 1998). Children of migrants – themselves born and/or raised in Europe, now adults with their own next generation(s) – often align with recognizable tropes about problems of integration as "citizens" and ethnicization as "other", in different modes across different states (Daoud, 2011).

One integral part of this integration-ethnicization dynamic is the potential to ignite a sense of heritage, family ties, and cultural identity by visiting "home". Even across distant generations, visiting places with ancestral connections engenders a sense of significance, belonging, and connection with others (Nash, 2008); but can also highlight distinctions, incompatibilities, and especially class differences in the form of mobile capitals (Coles & Timothy, 2004; Potter et al., 2005). The frequency and intensity of visits by diasporic European Moroccans to Morocco make this case distinct from many other migration trajectories, inasmuch as mobility between these continents is facilitated by relatively short distances, traversable by car and ferry, and increasingly served by low-cost airlines. Moreover, there have not been, like with some North African neighbors, significant periods of political unrest in Morocco to interrupt their flow. Yet, these mechanisms of facilitation also may increase their distinctiveness from resident Moroccan citizens, whose perceptions about travel and migration to Europe can be significantly colored (whether hopefully or discouragingly) through their networked contact with Moroccan-origin families in Europe (Jolivet, 2015).

Furthermore, over the years different efforts have been made to smooth the journey of *Marocains Résidants à l'Etranger* (Moroccans resident abroad) on their annual circulation between Europe and Morocco and attract them as leisure consumers in Morocco. They are literally and discursively welcomed each year by *Opération Marhaba* (Operation Welcome), an extension of previous transit service operations to facilitate vehicular traffic from Spain, rebranded into a "humanitarian action" (http://www.fm5.ma/en/marhaba) which involves nearly all government ministries providing services to diasporic arrivals. Since 2002, they also find a plethora of summertime "*festivals des plages*" by the state-owned Maroc Telecom (Groupe Maroc Telecom, n.d.), providing free high-profile concerts at many urban beaches, as well as concerts and events in central squares of inland cities. In short, the Moroccan state orients itself in multiple ways towards valuing these visitors by enthusiastically welcoming them and providing them with motivations to return through the service, excitement, and leisure promised in this homeland, rendering them particularly visible as a collective arriving and departing every summer.

These concerts might be read through the same strategies that cities elsewhere have used to engender and control "preferred" forms of leisure consumption, particularly for youth, in their nighttime economies. Such city development strategies have been characterized through a neoliberal lens, including an implicit tilt towards an intended "up-market" consumer seeking branded, and surveilled, experiential leisure (Chatterton & Hollands, 2002). Alcoholic drinking practices are often a focal issue for urban nightlife as experiential leisure in Western contexts (Latham, 2003; Roberts, 2015), along with attempts at the control of embodied practices and production of socially acceptable experiential leisure of excessiveness, release, and excitement through urban nightlife scenes (Hubbard, 2005; Jayne, Valentine, & Holloway, 2010; Malbon, 1999). Yet, for Muslim communities in European settings, businesses and atmospheres oriented around alcohol consumption can become sites of visceral exclusion (Valentine, Holloway, & Jayne, 2010), alongside other ways that ethnicized minorities may be characterized as "dangerous" andexcluded from nightlife centers (Schwanen et al., 2012). These exclusionary practices prompt a search for – or the creation of (Boogaarts, 2008) – alternative nighttime economies which are accessible to a minority consumer who seeks affects of experiential consumption parallel to that created by an inebriated "big night out" (Roberts, 2015). Unsurprisingly, then, part of what post-migrant generation diasporic visitors seek during their visits is unhindered participation in comparable "up-market" urban leisure centers in Morocco, where they are neither excluded by their religious preferences nor as a stigmatized minority.

This emergent combination of a state that is oriented towards producing cities as sites for tourism consumption for both foreign and diasporic tourists, along with a diasporic population who has the financial means and access to mobilities to visit relatively frequently, and the desire to find safe, friendly spaces for nighttime leisure, becomes a set of specific contextual conditions (Latham, 2003) for generating urban economies of visible, distinctively elite diasporic leisure. Without necessarily being aware of the processes that perpetuate it – or necessarily purposeful in their own actions and statements that engender classed distinction between themselves and resident Moroccans (Wagner, 2015) – these next-generation DVs arguably become, in their experiential leisure practices, contingent elites.

Analyzing affective practices

Empirically attending to formulations of emergence requires taking into account material and social becomings and the paths they depend upon; the modes through which agents interact with one another; and, most importantly, how all of these feed back into regeneration and differentiation (De Landa, 2002). By focusing in this example of how diasporic visitors become contextually elite in Moroccan urban spaces, this paper, as part of a larger project investigating diasporic belonging in Morocco (Wagner, 2011), reflects how eliteness might unexpectedly emerge in such an ecology in the dynamics of many social and material agents in interaction. In fact, eliteness was not a primary thematic concern of the original project, which was an attempt to ethnographically and ethnomethodologically comprehend the communicative and leisure practices of diasporic visitors during their summer holidays. That is, this analysis is built primarily upon ethnographic observation following mobile participants from three families, over several years as well as dozens of others encountered during fieldwork over two summers, with ethnomethodological attention to the practical accomplishment of visiting Morocco in concert with diasporic networks.

Effectively, eliteness only became relevant during analysis as more and more of the observed practices and reported discourses pulled me in to a social organization of economic, material and spatial boundaries differentiating DVs from local residents. As an ethnographer developing relationships with research participants, I also participated in an "education by attention" (Ingold, 2014, p. 388), learning instinctively how to follow practices and respond to rules that interlace to make up a collectivity. I learned to recognize what characteristics of distinction separate those who make (and can exceed) rules from those who do not. Close participant observation brought attention to *how* such rules operate to keep some members in and others out, and how they can repeat and differentiate with feedback. In that sense, this ethnography is ethnomethodological by focusing on the interactional affiliations and boundaries emerging in the micro-activity of everyday life (Goffman, 1971), and in this paper particularly on micro-activity and materiality of mobility and encounter in urban public space (Jensen, 2006; Wilson, 2016). The data presented below, as well as my analysis, is therefore based not only on practices observed, but also on how I oriented to them as an "educated" ethnographer along the rules and mobilities I learned to follow with DVs and in Moroccan public space more generally.

Discursively, the premise of contingent eliteness as outlined at the beginning of this paper is built on a perception of a perception as recounted in an interview context. Conversations about visitation practices were also collected in this project, and the metaphors and frameworks participants used to describe their orientations towards those activities are taken literally and seriously in relation to ethnographic observations (Katz, 1999). These meta-discussions often indicated attitudes that might be difficult to observe as "practiced" – such as Soumia's possible choice *not* to visit Morocco, which she may attribute (at least in part) to feeling annoyed at being perceived as arrogant. Such reports also indicate some idea of intentionality: that the production of eliteness may be, at least in Soumia's opinion, undeserved, though she may also recognize that others of her peers are acting "arrogant".

Soumia's comments also put into evidence a relevant background for discursive perceptions feeding back into each other, which may materially participate in how diasporic visitors orient towards others in Morocco. A potential awareness of a circulating "arrogance", though it may not be explicit in any particular observed encounter, may be acting as part of their cumulative and collective, affectively practiced orientation towards the holiday. That is, Soumia does not report an event of being told explicitly that she is arrogant; rather she is implicitly aware of that distinction characterizing interactions in Morocco of others like her as a general collective, and therefore possibly she herself as well. Furthermore, rather than being a direct consequence of any one moment of interaction – e.g. as explicitly linked to something Soumia did or said that makes others perceive her as arrogant – it behaves as an accumulation of the behavior of many bodies (Saldanha, 2007) It takes on its own momentum in characterizing these individuals into groups and construing perceptions themselves as a fuzzy group, and as an "other". As much as this analysis focuses on the embodied practices of DVs, it is essential to acknowledge that observed events may also be sites of affective practice extending beyond the immediate circumstance (Rutherford, 2016).

In order to attempt to integrate the presence of such cumulative, backgrounded orientations, I draw upon Wetherell's (2015) methodological directive to focus on the social practices of affect rather than on their nebulous and elusive "circulation". Affect as social practice is, effectively, a pre-condition of the fieldwork example below, describing an evening out with a group of diasporic visitor friends through their consumption movements around the city. This event is predicated on the search for the "big night out", itself an affective project of experiential leisure, whereby friend groups follow trajectories of movement and consumption over the course of an evening. As I have elsewhere devoted closer attention to how these spontaneous mobilities become an objective in themselves for how to spend leisure time (Wagner, 2017), the analysis here focuses more on the affective collaborations and distanciations as they occurred between this group and others they encountered along the way. Rather than attempting to define what this affect "is", I take it, ethnomethodologically, as a black box that becomes empirically relevant to the sequence of activity through its pushing and pulling of practices.

Using Wetherell's notion of affective practice as "a figuration where body possibilities and routines become recruited or entangled together with meaning-making and with other social and material figurations" (2012, p. 19), the fieldnotes excerpted below describe these social-material figurations through both what I observed participants doing during this "big night out", as well as my own responses and interpretations as an attentive ethnographer, participating in and contributing to these affective practices myself. I take these edited notes as data, and reproduce them here in a raw form, because both their form and content contributes to an analysis of affective practices that occurred on this night out to the extent that my own "education" as ethnographer is evident in how I frame what I (am able to) observe (Wolfinger, 2002). They indicate how affective practices contribute to the experiential objective of finding spontaneous fun around this city, and how these practices engaged multiple agents, including the group I was following, myself and others who were bystanders or contributors to our pursuit – like service staff, other drivers on the same streets, other leisure-seekers and family members staying at home on this summertime Wednesday.

Keeping in mind the production of "arrogance" as introduced by Soumia as potentially always relevant (or omnirelevant; Fitzgerald, Housley, & Butler, 2009) to how these interactions take place, this compilation of notes becomes as much an effort to understand how these friends organize their "night out" together in pursuit of an elusive affective state, as how others they encounter might read their behavior and engagement as "arrogant" through an affect of eliteness. Taking their mobility as a starting point for how they are carving out "elite" spaces from ordinary circulations, three significant mechanisms for distinction become visible in this evening, as I encoded into the fieldnotes in these figures: (1) financial accessibility for *consumption*, in how their access to capital makes them able to participate in specific environments; (2) bounded patterns of *circulation*, related both to their financial position as diasporic visitors and their purposes and practices as comparatively elite leisure consumers; and (3) being able to make and *exceed* rules of behavior in the different environments where they spend time. These mechanisms frequently intersect and overlap, as indicated in the coding, but coalesce in ways that reinforce how this group might be perceivable as "arrogant".

A Wednesday night in Fez

Fieldnote extract 1

> {circulation} Noura, who has a beautiful and captivating, often well-made up face, and her bande des amis, which is Kamal her neighbor; Hind, Fatiha and Lubna, who is with Mohammed; plus Karim the algerian, his friend Nabil, and Abdel and Norddine. {/circulation}
>
> {circulation} So much conversation that it's difficult to remember any single specific part but the overall impression was like high school summer weekends, where everyone has a car and the group clique feeling is so strong that you just want to keep performing well enough to stick around. {consumption} Moving from cafe to cafe, just to keep ourselves entertained, stay out a little later, stay with each other a little longer. someone says to me, on est en vacances.{/consumption} {/circulation}

Dating from August 2008, these fieldnotes document one group's leisure pursuits during fieldwork in Morocco. As the Fieldnote Extract 1 describes, this group was a band of eclectic friends: not family members or even from the same hometowns in France, but collected together through chance encounters and networked connections over their years of annual visits to Morocco (Fieldnote Extract 2 below). They were all in their late twenties to mid-thirties, mostly single (except the one couple mentioned), and were visiting Morocco as independent adults – that is, not necessarily accompanied by or accompanying their families, though they may spend time with family members during the trip. Some of them had family in Fez (such as Noura's mother, waiting for her at home) where this evening took place, while others had families in other towns and cities and were stopping in Fez for vacation.

At the point when I met them in this summer, they had already spent some time (from a few nights to a few weeks) on their holidays in Fez, though for the most part they were moving around Morocco (in smaller cohorts) and going out in different cities. Already, this combination of incidental encounters that spawned these friendships, as well as their uncoordinated and coincidental co-presence at the same time and in the same city in Morocco, feed in to the spontaneous affective sensibility of a "big night out" like this one as full of possibilities as they move around the city. Their ability to meet each other, as much as the story recounted here of this night out, indexes as well the bounded *circulation* of diasporic consumption spaces in Morocco, facilitating encounters between this mobile group as they might find each other in this or other cities within a semi-enclosed spaces like the various cafés, restaurants, and cars in which we spent most of the evening (Figure 1 and Figure 2).

Fieldnote extract 2

> {consumption} All the places we go (except the one I don't see, at Atlas) have a flavor of upperclassness in comparison to other similar cafes. The one we don't go to, Eight, has the

> most fanciful appearance; where we start, La Villa, is a bit Venezia Ice ish, lots of marble and steel and glass.{/consumption}
>
> {circulation}{consumption} La Villa is also a car place – far enough distant from walking sort of neighborhoods that you wouldn't happen upon it – it must be intentional, and it seems to be surrounded by villas. I think Noura's house is not far from it.{/consumption} {/circulation}
>
> {circulation} I ask how they all know each other – links seem to be more in Morocco than france. one is friend of another, or met at the pool a few years ago, or met this year for the first time. I get the sense that they contact each other on site, not really keep in touch outside of Morocco, especially since they all live somewhat distant except for individual pairs. {/circulation}
>
> {circulation} sitting there, Abdel & Norddine arrive with another girl who disappears soon after. {/circulation} {circulation} Long conversation with Norddine about research – seems interested in it. we discuss his travel this year – with parents – and meeting up with Abdel who drove alone.{/circulation}

Much like the story of this night out, their presence in Morocco was characterized by movement: they were moving around the country as they moved around during this evening, from place to place in a sea of nighttime leisure consumption options. Starting from where they originally met up with one another for this night out (Figure 2), their movements around the city that evening frequently intersected with other DVs who met us briefly, or joined in our trajectory (Fieldnote Extract 2). These movements created a bounded circulatory space, where they are able to encounter others like them, but as it involves moving from place to place through urban public spaces, they are not by any means invisible. Rather, their movements – and affective practices of group behavior, such as the argument that happened on the street – spill out into public view. That argument stretched across two separate consumption spaces, as it began at La Villa and continued as we made our way to Verdi.

Fieldnote extract 3

> {consumption}{circulation} hunger is agreed, and we get up to leave la villa but are stalled by the business between Nordine and Fatiha. Karim is so sick of waiting that he jumps in a taxi and goes ahead of us to Eight. hanging around the cars, waiting, until they finish suddenly and F insists we go to Verdi, she's sick of Eight {/circulation} {/consumption}
>
> {exceeding} Noura explains when we get there that they go to Eight quite often but the menu is limited and gets boring after a while. By now we're about midnight, 1am. {/exceeding}
>
> {circulation}Outside of Verdi, F calls Karim, who is purportedly just down the street, to come meet us.{/circulation}
>
> {exceeding}lots of migrating from table to table, outside to inside to banquette, to chairs. The waiter has a look on his face like he's really not interested in getting this business, and we take a long time to organize and order. Only Lubna is consistently speaking arabic – she tells jokes and stories, she makes polite concessions to the waiter.{/exceeding}

In many ways, the evening documented in this story can be characterized as "youth culture", and not necessarily "elite". In writing my impressions in Fieldnote Extract 3, I compared it to youthful summer weekends spent hanging out in a clique of friends, simply trying to stay out late with each other. Yet, the places that were embedded in this group's pattern of circulation – several cafes which I commented on as having a "flavor of upperclassness" (Fieldnote Extract 2) – mark their circulations as financially distinctive in a landscape of ordinary consumption in Morocco. While these do not have the same high bar for entry as other potential spaces for consumption in Morocco that might be only accessible to a "superrich" clientele, they represent a cumulatively expensive experience – tallying about 30 to 40 Euro per night (Fieldnote Extract 5), per person in a country where the standard minimum salary is about 300–400 Euro per month. Their "being on vacation" (Fieldnote Extract 1) combined with their trajectories as diasporic visitors with access to European cost-of-living capital makes these locations accessible to them as comparatively "cheap" for consumption. Furthermore, their circulations were in parallel with other, similarly configured diasporic groups, like the "Belgians" who kept La Villa cafe open late alongside us (Figure 4). Their circulations, though they may seem erratic, are following specific pathways and limited to certain corridors of urban space (Figure 3).

Fieldnote extract 4

> {exceeding}{circulation}leaving Verdi becomes a car parade – sort of following a wedding parade that just passed us, we head off toward 'Atlas'. there are car to car conversations: Noura feels like un peu de la marche [a little walk], after eating, so she pulls up alongside one vehicle then the other to suggest it (on Blvd Hassan II, which is not totally empty but mostly so). as Mohammed and Lubna's car passes us a bit further up, L shouts 'pas de la marche!' [no walking!] out the window.{/circulation}{/exceeding}
>
> {exceeding}{consumption}Atlas turns out to be a totally unexpected dirty coffee shop next to the CTM station. Unfortunately, it's closed until 3am. {/consumption}{/exceeding} {circulation}At that point, one car had gone off to pick up F's jacket, so we must be communicating with them by mobile to tell them where to end up, or tell us where to go. N's car follows M&L back to La Villa, where they are waiting.{/circulation}

As consumption choices, these locations are contingent to their mobility practices, perceptions about the urban environment (Figure 4), and their social networks. Though following their movements linearly may make them seem random – spontaneously bouncing from one location to another without any specific planned destination, simply responding to a capricious desire of one member (hunger, walking versus driving, forgotten jackets) – when followed along night after night, their bounded and systematic circulation becomes more evident. The pattern is chaotic as opposed to random, starting from meeting in Morocco as a more-or-less cohesive group of friends, all with the same purpose of "being on vacation" (Fieldnote Extract 1), and moving between a range of possible consumption activities for accomplishing that practice, located in a limited range of spaces that meet a limited scope of needs for their purposes. These needs are partially coordinated through socially configured dimensions pertinent to Morocco as their "homeland" destination which I have discussed elsewhere, such as having access to a car (Wagner, 2017) or feeling "safe" in mixed-gendered groups (Wagner & Peters, 2014). Both of these dimensions are path-dependent on class-flavored distinctions, in that both access to a car (and the concomitant need for a place to park – see Figures 1 and 2) and "safe" consumption spaces for women outside the home – such as cafés and restaurants that are not entirely male-dominated, or nightclubs where women can avoid being mistaken for prostitutes (see Graiouid, 2007; Wagner & Peters, 2014) – hinge upon markets for these spaces that are contingent upon the presence of enough consumers for them. In many ways, diasporic visitors contribute to the existence of that market in Morocco, to the extent that they constitute a significant number of paying customers who will reliably return every summer, but they are also embedded in emerging gentrification and cosmopolitan consumption scenes that serve elite client bases in Morocco. In this sense, they are creating these consumption spaces through their presence, while simultaneously animating them with their affective practices of seeking fun in ways that are specifically configured to their diasporic context.

Fieldnote extract 5

{consumption} we are a sort of energetic force here again: F orders a box full of cookies and sweets, and a bunch of us order desserts, nearly cleaning out the remaining gateaux from the cold case. {/consumption} {exceeding} The waiter here looks more happy to see

Figure 1. A cafe popular with DV participants (Photo by author).

Figure 2. La Villa café, photographed after these events (Photo by author).

> us, but not completely – it seems like he's allowed to go home as soon as all these customers are finished. In fact, the don't really force us out but once everyone has finished the cafe suddenly shuts down. {circulation} (another car full of belgians was there at the end. . .) {/circulation}{/exceeding}
>
> {circulation}{consumption} Conversation here is a little bit slower – seems like some of us are getting tired. At one point Mourad (?) shows up – Moh's brother? – and talks about where he's been that night: all the nightclubs in Fes, one after the other. {/consumption}{/circulation}
>
> {consumption} It occurs to me as we pay the bill, approx 30 euro for everyone, that none of these excursions are individually very expensive, but they add up to probably 30-40 euro/day for various meals, plus poss. car rental at another 30. all things considered, this is not very expensive for a vacation, but it's still a lot more than a local moroccan would be able to afford, even on holiday, {/consumption}

Furthermore, in inhabiting these consumption spaces and making them their own, the above notes indicate, indirectly, how these participants become powerful enough in these environments to make and/or break rules. While often these were not explicit or necessarily enforced rules, they were rules for social ordering that garnered an affective response when they had been exceeded. During this evening's activities, I describe how they entered restaurants and "energized" them, taking over or rearranging several tables and expecting service beyond normal business hours (Fieldnote Extract 5). These become scenarios where they make their own rules, taking over spaces without requesting or securing permission (Fieldnote Extract 3). At other moments they were interactionally exceeding their roles as "customers", marked by staff members telling them the kitchen is closed or suddenly shutting down a restaurant where they have arrived after (or before) the normal service hours (Fieldnote Extract 5). These microbehaviors of politeness (Culpeper, 2011), especially in service encounters (cf. Hochschild, 1983) – marked here as well by Lubna's overt and

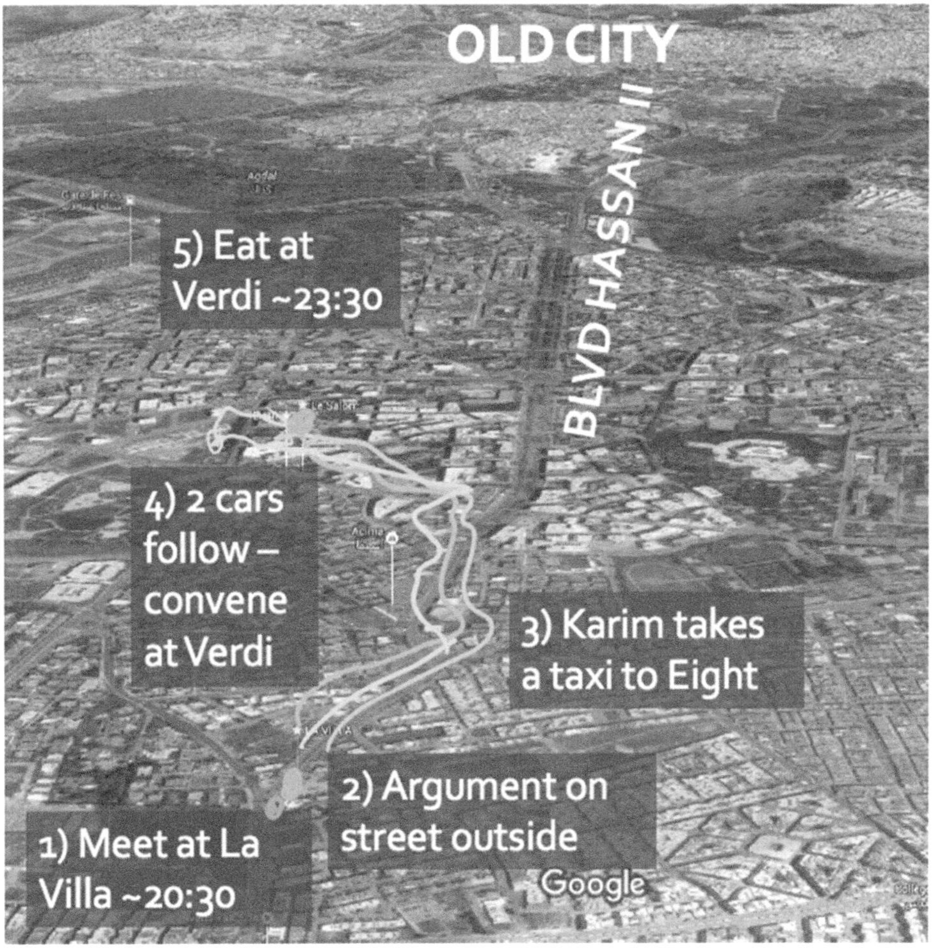

Figure 3. Annotated map of Fez, looking up Boulevard Hassan II (Google Maps, Author's annotations).

noticeable efforts to play and joke in Arabic with the disinterested waiter (Fieldnote Extract 3) – is exactly the affective practice that might engender the perception of arrogance around their activities. Their excessivity was also marked by my own discomfort and questions about their behavior, in my noticing their late-night erratic driving (Figure 4, Fieldnote Extract 4) as they pulled up alongside each other on a central boulevard which still, at that late hour, had some rule-following traffic moving alongside us, and in my question to Noura about her mother's approval for being out late at night (Fieldnote Extract 6). Each of these indicates assumptions about rules – rules of cafe-going, rules of traffic flow, and rules of gendered public behavior – that this group was fluidly and enjoyably pushing to the point of violation in their urban leisure practices. These affectively exhilarating rule-breaking practices might compare favorably to affective experiences they, or others of their cohort, might have in trying to access European nighttime activities, as discussed above.

Figure 4. Annotated map of Fez, extended view. (Google Maps, Author's annotations).

Fieldnote extract 6

> {exceeding} I asked Noura at some point if her mother is ok with her being out late, and she says she [mother] is divorced, and she trusts her; however, when heading home, she wakes up Kamal – is this because she's nervous about the walk from her car to her house? or because she wants to look like she has her guard of honor? {/exceeding}

As much as these affective practices exemplify youthful leisure in their breadth of accessibility and energy for excess, they also reach an apex. As chaotic patterns, they have limiting points at which their patterning shifts suddenly into different modes of repetition. The final comment on Noura's desire for an escort to her house (Fieldnote Extract 6) is one example of this: though she has previously told me her mother "trusts her" to go out late as a single woman, she still feels a need for gendered protection to arrive home. In fact, similar protection was enforced on me, as the members of the group who drove me home insisted on accompanying me until my door (in the old city – a residential and consumption space we had not visited on this night out). We were both effectively departing from our elite bounded consumption space and the cars that carried us between them, rendering both of us potentially vulnerable to unknowns beyond this group and the effectively enclosed spaces it occupies.

Feedback: eliteness of mobile experiential consumption

This series of explorations, retracings, and feedbacks come to make up an example of affective practices of eliteness for DVs in Moroccan cities during summer holidays. While the individuals described above – and the many others who engaged in similar frenetic mobilities of leisure consumption during the fieldwork period (see Wagner, 2011) – are by no means globally elite, they manage to achieve some kind of elite differentiation across urban landscapes through their search for experiential leisure. This eliteness is in some ways materially dependent upon their own genetic and migratory lineages, as descendants of Moroccan parents who chose mobility in search of economic advantage and succeeded, and whose annual visits from youth to adulthood to their ancestral homeland have trained them to explore it as site for pursuing vacation leisure. Though the example above documents single adults, similarly chaotic, mobile, experiential-leisure-seeking was a feature of many different cohorts of Moroccan DVs, from youth to parents, from individuals to families. These leisure pursuits are also dependent on the existence of consumption sites like those consumed as described above, which offer services and experiences that correspond to their desires for non-discriminatory leisure experiences akin to those on offer in their European homes. In many ways, these fit the "touristic" orientation of Morocco towards a European consumer (Minca & Wagner, 2016). As much as they might have many non-touristic reasons for visiting Morocco more directly and discursively defined by their genetic lineages – such as the desire to "know one's roots" by maintaining ties with family and being familiar and "at home" in a homeland – these repeated and iterated practices become an affective motivation as well.

Though the events of this evening were confined to one city, these practices in other cases might roam into other cities around Morocco and back, within a single night out. In parallel with Roberts' interviewees practicing experiential leisure in the UK, DVs' search for leisure sites is not limited to what is nearby, but extends to what might provide the spontaneity of a new and memorable experience (2015, p. 576). These hypermobile practices are themselves a form of inaccessibility: they render these groups contingently elite through their consumption-enabled movement, as their leisure landscapes extend physically beyond and propel speedily past what is normally feasible for resident Moroccans around them. DVs become, visibly and practically, the "conspicuous consumer" in contrast with resident others with whom they come into contact, and possibly engendering the perception of "arrogance" that Soumia reported simply by joining along with their friends. This "arrogant" eliteness is thus, in some ways, inherent in their affectively practiced presence: like Stephenson (2002) describes, the visiting friends and relatives to migration-sending countries can often be considered as elite – against their wishes – simply through their implicit access to mobility of self and economic capital, in contrast to their loved ones. They are "on vacation", spending money, and seeking experiences that often exceed the financial habits of residents who they might know or encounter.

Beyond these distinctions related to migratory mobility, their leisure-seeking is also part of the rule-making (and rule-breaking) power of being the group who defines what is "in" in urban leisure scenes (Saldanha, 2007). As such, this becomes a complex desirable and undesirable eliteness, whereby DVs may recognize themselves as being (unintentionally) classified as elite in their access to capital, but willingly seeking to participate in ongoing leisure that perpetuates that categorization. Some of the behavior detailed in the fieldnotes above indicate a sense of Morocco as a place where rules do not apply – neither to official actors nor to themselves, as they learn to "get away" with transgressive behavior, like driving erratically (Figure 4, Fieldnote Extract 4) because officials can be paid off to ignore it. Such rule-breaking does not need to be officially consequential; as it becomes an accepted mode of affective practices for "being on vacation", it can expand and extend by reproducing in other activities, like when the economic presence of this group (and another Belgian cohort) keep a café open past their normal hours. Yet, as much as this rule-exceeding or -breaking is accepted within this cohort, it is still marked by others as unacceptable in their reactions to it – reactions which are sometimes recognized and responded to, and sometimes unnoticed. This power as a rule-breaking elite becomes similarly inadvertent as their conspicuous consumption, but also equally consequential in defining distinctive borders between DVs and resident Moroccans and engendering a divisive sense of "arrogance".

Conclusion

Whether the eliteness observed in these dynamics is purposefully "arrogant" or inadvertently emergent, it contributes to the making of Moroccan cities as they materially contour along with the practices of diasporic visitors. Envisioning how this night out repeats for many young DVs over many summer nights and across many Moroccan cities, we can extrapolate the accumulating economic pressure for urban nighttime consumption environments configured to serve these ready consumers – effectively the archetype of "up-market" leisure consumer the Moroccan state is orienting towards in its projects for welcoming Moroccans from abroad. The relatively bounded circuit of spaces where they hang out and encounter each other – evident as critical masses of many bodies even in such small as moments when this group crossed paths with others engaged in parallel leisure trajectories – become entangled in both their affective practices of visiting Morocco and a practiced, unintentional eliteness.

The perception of "arrogance" that Soumia identified at the introduction becomes empirically parseable through the combination of their circulation, consumption and rule-bending practices. As some move through the city with vehicular speed, from consumption site to consumption site – even returning to repeat a consumption site in the same evening – they are displaying access to economic power that distinguishes them from the local residents whom they might encounter. Likewise, their fun-seeking behavior demonstrates a disregard for social order that would normally structure consumption environments (like appropriate times to order food at a restaurant) or mobile circulation (like rule-following practices for driving in city streets). One of these dynamics alone might not be significant enough to mark out this group as "arrogant", but their combination becomes more than the sum of its parts, accumulating into affective practices of elite distinction that become generically attributable across the many different diasporic visitors present in Morocco. This eliteness is not precisely measurable as differentials of wealth or income, or their access to cars; moreover, this eliteness is not, as noted in observations of other kinds of elites, invisible. Rather, it is the metrics of differential consumption style and speed, combined with the affective ways they become remarkable and potently visible in social spaces via repeated practices of exceeding social order out in the city.

As affective practices of experiential leisure that pull DV participants along with it – triggering the departure to Morocco for many, and characterizing how leisure time is spent while there – these practices accumulate beyond their immediate goals for going to Morocco, and can become an affective goal in themselves. The ways their nighttime leisure involves taking up space and reorganizing its social rules along its logics – effectively changing rules and defining new boundaries – renders these individuals, in their participation in these affective practices, contingently and emergently elite.

Acknowledgments

Thanks are due to all of the participants in this project for allowing me to stay out all night with them. I send thanks as well to the anonymous reviewers for their extremely detailed and useful comments on the first version of this paper, to David Bassens and Bas van Heur as session organizers and special issue editors, and finally to Alan Latham as continuing consigliere.

Disclosure statement

No potential conflict of interest was reported by the author.

Funding

This work was supported by the University of London Central Research Fund.

ORCID

Lauren B. Wagner http://orcid.org/0000-0002-4778-7408

References

Bassens and Van Heur. (2019). An urban studies approach to elites: nurturing conceptual rigor and methodological pluralism. *Urban Geography*, *40*(5), 591–603.

Boogaarts, Simone. (2008). Claiming your place at night: Turkish dance parties in the netherlands. *Journal of Ethnic and Migration Studies*, *34*(8), 1283–1300.

Chatterton, Paul, & Hollands, Robert. (2002). Theorising urban playscapes: Producing, regulating and consuming youthful nightlife city spaces. *Urban Studies*, *39*(1), 95–116.

Chattou, Zoubir. (1998). *Migrations Marocaines en Europe: Le Paradox des Itinéraires*. Paris: L'Harmattan.

Coles, Tim, & Timothy, Dallen J (Eds.). (2004). *Tourism, diasporas and space*. London: Routledge.

Culpeper, Jonathan. (2011). 'It's not what you said, it's how you said it!' Prosody and impoliteness. In Linguistic politeness research group (Ed.), *Discursive approaches to politeness* (pp. 57–83). Berlin, Boston: De Gruyter.

Daoud, Zakya. (2011). *La Diaspora Marocain En Europe*. Casablanca: La Croisée des Chemins.

De Landa, Manuel. (2002). *Intensive science and virtual philosophy*. London: Continuum.

Fitzgerald, Richard, Housley, William, & Butler, Carly W. (2009). Omnirelevance and interactional context. *Australian Journal of Communication*, *36*(3), 45–64.

Freudendal-Pedersen, Malene. (2014). Tracing the super-rich and their mobilities in a scandanavian welfare state. In Thomas Birtchnell & Javier Caletrío (Eds.), *Elite mobilities* (pp. 209–225). Abingdon: Routledge.

Goffman, Erving. (1971). *Relations in public: Microstudies of the public order*. New York, NY: Basic Books.

Graiouid, Said. (2007). A place on the terrace: Café Culture and the public sphere in morocco. *The Journal of North African Studies*, *12*(4), 531–550.

Groupe Maroc Telecom, (n.d.). Le festival des plages, de Maroc Telecom. Retrieved February 27, 2017, from http://www.iam.ma/groupe-maroc-telecom/responsabilite-societale-dev-durable/nos-realisations/promouvoir-la-culture-et-les%20arts/le-festival-des-plages-de-maroc-telecom.aspx

Hillali, Mimoun. (2007). *La Politique du Tourisme au Maroc: Diagnostic, Bilan et Critique*. Paris: L'Harmattan.

Hochschild, Arlie. (1983). *The managed heart: Commercialization of human feeling*. Berkeley: University of California Press.

Hubbard, Phil. (2005). The geographies of 'going out': Emotion and embodiment in the evening economy. In Joyce Davidson, Liz Bondi, & Mick Smith (Eds.), *Emotional geographies* (pp. 117–134). Aldershot: Ashgate.

Ingold, Tim. (2014). "That's enough about ethnography! *HAU: Journal of Ethnographic Theory, 4* (1), 383–395.

Jansson, André. (2016). Mobile elites: Understanding the ambiguous lifeworlds of sojourners, dwellers and homecomers. *European Journal of Cultural Studies, 19*(5), 421–434.

Jayne, Mark, Valentine, Gill, & Holloway, Sarah L. (2010). Emotional, embodied and affective geographies of alcohol, drinking and drunkenness. *Transactions of the Institute of British Geographers, 35*(4), 540–554.

Jensen, Ole B. (2006). 'Facework', flow and the city: Simmel, goffman, and mobility in the contemporary city. *Mobilities, 1*(2), 143–165.

Jolivet, D. (2015). Times of uncertainty in Europe: Migration feedback loops in four Moroccan regions. *The Journal of North African Studies, 20*(4), 553–572.

Katz, Jack. (1999). *How emotions work*. Chicago: University of Chicago Press.

Latham, Alan. (2003). Urbanity, lifestyle and making sense of the new urban cultural economy: Notes from Auckland, New Zealand. *Urban Studies, 40*(9), 1699–1724.

Malbon, Ben. (1999). *Clubbing: Dancing, ecstasy and vitality*. London: Routledge.

Michael, Hall, C., & Tucker, Hazel (Eds.). (2004). *Tourism and postcolonialism: Contested discourses, identities and representations*. London: Routledge.

Minca, Claudio, & Wagner, Lauren. (2016). *Moroccan dreams: Orientalist myth, colonial legacy*. London: I.B. Tauris.

Myers-Scotton, Carol. (1990). Elite closure as boundary maintenance: The case of Africa. In Brian Westein (Eds.), *Language policy and political development* (pp. 25–42). Norwood, NJ: Ablex.

Nash, Catherine. (2008). *Of Irish descent: Origin stories, genealogy, and the politics of belonging*. Syracuse: Syracuse University Press.

Potter, Robert B., Conway, Dennis, & Phillips, Joan (Eds.). (2005). *The experience of return migration: Caribbean perspectives*. Aldershot: Ashgate.

Roberts, Marion. (2015). 'A big night out': young people's drinking, social practice and spatial experience in the 'liminoid' zones of english night-time cities. *Urban Studies, 52*(3), 571–588.

Rutherford, Danilyn. (2016). Affect theory and the empirical. *Annual Review of Anthropology, 45* (1), 285–300.

Saldanha, Arun. (2007). *Psychedelic white*. Minneapolis: University of Minnesota Press.

Schwanen, Tim, van Aalst, Irina, Brands, Jelle, & Timan, Tjerk. (2012). Rhythms of the night: Spatiotemporal inequalities in the nighttime economy. *Environment and Planning A, 44*(9), 2064–2085.

Spence, Emma. (2016). Eye-spy wealth: Cultural capital and "knowing luxury" in the identification of and engagement with the superrich. *Annals of Leisure Research, 19*(3), 314–328.

Stafford, Jean, Bélanger, Charles-Etienne Sarrasin, & Sarrasin, Bruno. (1996). *Développement et Tourisme au Maroc*. Montréal: Harmattan.

Stephenson, Marcus L. (2002). Travelling to the ancestral homelands: The aspirations and experiences of a UK caribbean community. *Current Issues in Tourism, 5*(5), 378–425.

Thurlow, Crispin, & Jaworski, Adam. (2012). Elite mobilities: The semiotic landscapes of luxury and privilege. *Social Semiotics, 22*(4), 487–516.

Valentine, Gill, Holloway, Sarah L, & Jayne, Mark. (2010). Contemporary cultures of abstinence and the nighttime economy: Muslim attitudes towards alcohol and the implications for social cohesion. *Environment and Planning A, 42*(1), 8–22.

Wagner, Lauren, (2011). *Negotiating diasporic mobilities and becomings: Interactions and practices of Europeans of Moroccan descent on holiday in Morocco* (Doctoral thesis). Geography, University College London, London. Retrieved from http://discovery.ucl.ac.uk/1317815/.

Wagner, Lauren. (2015). Shopping for diasporic belonging: Being "local" or being "mobile" as a VFR visitor in the ancestral homeland. *Population, Space and Place*, *21*(7), 654–668.

Wagner, Lauren. (2017). Viscous automobilities: Diasporic practices and vehicular assemblages of visiting 'home'. *Mobilities*, 1–20. (Online)

Wagner, Lauren, & Peters, Karin. (2014). Feeling at home in public: Diasporic moroccan women negotiating leisure in morocco and the Netherlands. *Gender, Place & Culture*, *21*(4), 415–430.

Wetherell, Margaret. (2012). *Affect and emotion: A new social science understanding*. London: Sage.

Wetherell, Margaret. (2015). Trends in the turn to affect: A social psychological critique. *Body & Society*, *21*(2), 139–166.

Wilson, Helen F. (2016). On geography and encounter: Bodies, borders, and difference. *Progress in Human Geography*. (online).

Wolfinger, Nicholas. (2002). On writing fieldnotes: Collection strategies and background expectancies. *Qualitative Research*, *2*(1), 85–93.

Investigating urban elites through fear and security in Porto Alegre

Fabio Vanin

ABSTRACT
Taking from the results of the ongoing EU funded research "Reducing Boundaries" (2014–2017), the paper contributes to the understanding of the 'urban geographies of elite reproduction" through the lens of the emerging "culture of fear" as a fundamental contemporary societal paradigm. Today, urban elites behaviours, lifestyles, discourses and their spatial organisation are in fact increasingly driven by real or perceived urban insecurity, with direct consequences on the built form and socio-spatial organisation of their cities. As a consequence, understanding the building types and urban patterns those groups are direct or indirectly proposing and their impact on society, searching for an effective methodology to conduct such investigations is at the core of the the research discussed here. The paper will examine the paradigmatic case of Porto Alegre showing urban elites' defensive practices and interpreting their impact on the urban realm.

Article

The interest in studying and trying to define the elites in different urban contexts derives from the political, economic, and socio-cultural power these groups can exert on society and the living environment. Stemming from the results of the EU funded research "Reducing Boundaries: Understanding Exclusion Through Security Defensive Systems" (2014–2017),[1] this paper contributes to the understanding of the "urban geographies of elite reproduction" through the lens of the emerging "culture of fear" as a fundamental contemporary societal paradigm (Bauman, 2003; Beck, 1992; Casati, 2012; Goldstein, 2010; Low, 2004). Today, the behaviors, lifestyles, discourses, and practices of urban elites are in fact increasingly driven by real or perceived insecurity, with direct effects on the built form and the socio-spatial organization of cities. Therefore, studying the architectural types and urban patterns these groups directly or indirectly propose–starting from the observation of (elitist) private spaces as some of the most influential spatial patterns–can reveal the existing strict relationship between elitist behaviors, security, and the construction of the group identity. Searching for an effective methodology to conduct such investigation is at the core of the above mentioned research.

This paper examines the case of Porto Alegre, a "second (or third)-tier city" that represents one of the most extreme examples[2] where the security paradigm plays a key role in driving urban development and where the lifestyle of the upper classes dominates the common imaginary. Within the Brazilian context, Porto Alegre represents a particular case of the polarization of wealth and shrinking use of public and shared spaces in the city, to a certain extent more extreme than other large urban contexts in the country.

Reflecting on contemporary elitist living conditions and standards in Porto Alegre seems particularly fruitful to outline emerging trends, such as the growing investment in defensive systems and in exclusive real estate projects, processes of formal-informal occupation and control of public spaces, and mechanisms of spatial reshaping, production, and reproduction. Moreover, they reveal tendencies that go far beyond the so-called Global South, shedding light on emerging European trends and raising a series of key questions about the transnational influence of urban models, about the references that urban elites are explicitly or implicitly suggesting, and about the mechanisms of segregation and exclusion that are being introduced.

This paper discusses the advantages of taking the issues of fear and security as fruitful starting points to understanding urban elites in Brazil and specifically in Porto Alegre (The Lens of Fear and Security), and highlighting the need for a contextual approach to reach a satisfactory definition (Toward a Definition of Porto Alegre's Urban Elite). Second, the paper describes the construction and application of a specific methodology at the crossroads between urbanism and anthropology (Applied Methodology), and some of the most relevant emerging issues that characterize Porto Alegre's urban elite (Emerging Issues).

The lens of fear and security

Is it possible to research urban elites through interpretative categories such as fear and security? To what extent are those lenses able to provide a clear conceptualization of the elites?

Fear participates in shaping a city's form according to the structure, the relational mechanisms, and the perception of the social groups that inhabit cities, contributing to a greater understanding of the sociopolitical and historical dynamics that lead to the centrality of (in)security in contemporary societies. Fear can be considered as a transnational feature of globalization and an expression of deeper anxieties associated with its modernizing project (Dammert & Malone, 2003) even where the possibility of being a victim of violence and the danger of the urban environment are strong realities.

The material expressions of the "fight against insecurity" (armed guards, walls and fences, enclosed neighborhoods, an army presence in public spaces), together with the emotional geography of fear and the symbolic dimensions that lead to certain behaviors, contribute to the definition of social groups, including the upper classes. In many contexts–such as Porto Alegre–in which notable social frictions, wealth and income inequalities, and high crime rates are present, the issue of urban safety becomes a dominant discourse that has an important impact on society (Vanin, 2015, p. 81). A number of authors have already highlighted how a large part of the manifestations of social disease is linked to distrust and an increased perception of insecurity and anxiety structuring everyday life, thereby generating an increasing demand for control and security measures. This in turn leads to the "rise of the security paradigm as a framework for organising contemporary social life" (Goldstein, 2010). Risk and (in)security have become central elements of contemporary life and key analytical references for the analysis of social structures, constituting the main features of what has been termed "late neoliberalism" (Lupton, 1999, 2013; Povinelli, 2011) and intrinsically defining the so-called "risk society" (Beck, 1992). With the loss of what Giddens calls "ontological safety" (1999), (in)security appears to be one of the main features of present-day societies, at a symbolic and at a material level.

Moreover, security is a clear indicator of strong socioeconomic inequalities in societies where a large proportion of the population are experiencing deep economic and social hardship. As a reaction, strategies aimed at guaranteeing security are used to prevent crime, but at the same time they contribute to processes of segregation and the exclusion of weaker groups (often associated with the origin of danger), leading to cultural differentiations, misconceptions, and distorted perceptions of "the other" (Casagrande, 2016).

The need for control and safety by the upper classes in an effort to manage uncertainty and fear has resulted in the development of different security measures that tend to put aside public freedom and human rights, which have become marginal and subordinated (Reiner, 2007). Private, shared, and public interventions against insecurity are progressively growing (Reid, 2003), while governance strategies mix security actions aimed at prevention with repressive treatment of safety emergencies (Bauman, 2003).

Having identified safety as a useful interpretative category to investigate urban elites, we must ask ourselves which indicators suggest their behaviors, discourses, and practices.

Viewing elites as unitary groups (class) relates to the adoption of physical security measures that strongly contribute to the construction of the "visual landscape of fear" (Low, 2001), which can be observed in cities such as Porto Alegre. In such a context, the effort of constructing a "secure society" in order to assure a "safe environment" is not only visible, but it appears as a common goal for the upper class, where risk management and control become (almost) the only possible principles that can guide the development of living spaces and lifestyles. Those measures tend to reproduce recurrent patterns in the physical and symbolic attempt at "designing out crime" (Coloquhoun, 2012; see also Newman, 1972), creating spaces that prevent offences by means of control and defense through specific high and low-tech devices (fences, walls, barbed and electrified wires, cameras, fingerprint recognition systems, etc.). The resulting type of cityscape, comprising privatized and highly controlled spaces, fenced and patrolled, is a direct consequence of individual initiatives, market offers, and media communications, which crystallize around a specific gated way of living. These factors play an important role in the transformation of the urban environment according to the paradigm of (in)security, shaping physical spaces and the everyday life of citizens, as well as influencing their daily routines, social relationships, and aesthetics.

Moreover, the fight against insecurity by the upper classes catalyzes against "violent crime," where "the criminal" is the object of fear: this seems to be a common trend in the way the issue of urban (in)security is addressed in different urban contexts, where the "fear of crime" has been used for the application of effective public policies (Dammert, 2012, pp. 27–28).

Thus, analyzing the discourses of the "war against crime" by dominant societies over recent decades allows one to understand their relationship with other social groups, their perceived threat of a bad "otherness," and their classification and segregation practices. The antagonism toward the objectified figure of "the criminal," "the poor," and "the immigrant" as dangerous and "alien to our daily life reality" (Dammert, 2012, p. 3) suggests that the upper classes incorporate them as symbols of fear and insecurity in direct association to social class and ethnicity (Casagrande, 2016). That symbolism highlights how often the perception of fear and criminal phenomena hide other types of insecurities, such as the mistrust of public authority, the absence of a welfare state, and the fear of losing individual or collective identity and social roots (Bauman, 2001; Dammert, 2012; Wacquant, 2006).

Examining the contextual, socio-cultural relationship between urban elites and other groups seen through the lenses of fear and security is therefore essential in order to reach precise interpretations. For example, in Brazil, such consolidated dynamics and diffused practices are present in various cities where processes of gentrification, suburbanization, and social polarization lead to urban contexts in which the displacement and control of the poor becomes functional to the security of the middle and upper classes. As underlined by Christian Kesteloot (2013) when discussing the "repressive city scenario," this negative vision exacerbates the tensions produced by the blending of the fear of the poor with the fear of the "other," and also constitutes a growing trend in Europe.

Toward a definition of Porto Alegre's urban elite

The diffuse presence of *condominios fechados* (the Brazilian version of gated communities) for the high and middle classes, the ongoing privatization and securitization of public and semi-public spaces, the polarization of wealthy neighborhoods, the forced evictions in poor areas, and the stigmatization of "porous" areas such as the city center (*Centro*) highlight trends that are in many ways common to other cities in the Global South. However, during the last 20 years in Porto Alegre, the absence of control and an urban vision by public authorities in favor of private developments, the massive mediatization of elite lifestyles, and the shrinkage of public-common facilities suggest the political and societal influence of an urban model that belongs to wealthier groups. Moreover, the high crime rates, the massive construction of defensive and control systems to secure private spaces, and the omnipresent discourse about security draw a direct line between the high and middle classes and fear with regard to the construction of the city.

For these reasons, during the two fieldwork projects carried out in Porto Alegre (in 2014 and 2016), reaching a precise and contextual definition of "the rich and upper class," through the lens of fear and security, was a mandatory step. The identification of that group was one of the goals of the research "Reducing Boundaries," which tried to answer the following question: what is the societal group that has the power to promote defensive housing models and to influence the real estate market and public decisions on planning processes?

That effort led to a transversal and more inclusive definition of "urban elites." To reach the definition, it was necessary to apply a multifaceted approach, taking into account the socioeconomic and spatial dynamics that have led to the present configuration of Porto Alegre–the capital of Rio Grande do Sul–its historical evolution, the observation of different households and daily behaviors, and the analysis of recurrent discourses on safety through ad hoc interviews.

Thus, the identification of local elites passed not only through a quantitative study but also a qualitative one. On the one hand we took into account statistical data on income and wealth and the governmental definition of "rich" and "super-rich," and on the other hand we conducted interviews with local experts in order to identify those groups that have the material as well as the socio-cultural and political means to be directly or indirectly considered part of the elite.

The research context and the specificity of Porto Alegre accordingly led to the inclusion of a broad group of individuals (our sample) that we can define as the "urban elite," who range from the economic category of "upper middle class" to a number of people who occupy a prominent role in the local society. This group escapes the "objective" definition deriving from statistical measurements, as we gave priority to the relevance of the issue of urban security and its impact on the built environment. Moreover, the quantitative approach presented, from the beginning, a series of problems.

Data taken from the Instituto Brasileiro de Geografia e Estatística (Brazilian Institute of Geography and Statistics, IBGE) census of 2010 shows the geographical distribution of the monthly average income in Porto Alegre, using relatively precise (small scale) statistical units. However, the advantages offered by the good spatial readability of income distribution does not fit in with the official definition of the threshold above which someone is considered as part of the upper class (*classe alta*: http://www.sae.gov.br/), that is, between approximately 2500 R and 6500 R (€800 to €2100) per month. That range is not only misleading and too low–due to it including an excessively varied number of different societal groups that are far from homogeneous–but it also drastically lowers the threshold of the super-rich (*classe alta-alta*: above 6500 R) representing the top of the pyramid. In fact, despite the national poverty rate in Brazil, in the urban context most of those whose income is above 6500 R should more realistically be considered as upper middle class. Moreover, the majority of the interviewees who declared earning more than that monthly average income were not aware of being part of the *classe alta-alta* and did not consider themselves as such.

Another aspect that calls for a contextual approach is the fact that Brazilian elites are strongly defined by the specific evolution of their society: since the end of the dictatorship (1985), among the wealthy civil society there has been a general growing distrust of public authorities (local government and the state), that are considered as incapable and unreliable. That diffuse opinion has led to self-organization processes by the upper class, with the objective of substituting the lack of power and control at the public level, resulting in specific actions, behaviors, built spaces, and lifestyles.

This again points to the need to go beyond a rigid definition of "economic elites" as an analytical category based simply on quantitative data regarding income and wealth. The huge discrepancies among the official Brazilian group of *classe alta-alta* highlight significant wealth differences within that category. In that regard, for Porto Alegre it was far more relevant to understand elitists' behaviors through qualitative analysis, which allowed us to identify "other elites" that should be included in the main category, such as those who mimic the super-rich lifestyles, perpetrating similar social and physical effects. As stated by Facchetti (2015, p. 37), "the description of wealth in Brazil cannot rely solely on quantitative parameters but has to resort to the analysis of the model of life." Following on from the work of the sociologist Cattani (2011), the case of Porto Alegre clearly shows the need to use both quantitative and qualitative analyses in order to comprehend and better define the Brazilian elites. Its upper middle class is characterized by multifaceted origins and characteristics, but follows similar consumption models based on property ownership and an ostentatious lifestyle. That description escapes the traditional definition of class–as a homogeneous group sharing similar attributes such as identity, unity, and cohesion–and strongly invites the application of a variety of methods to uncover urban elite geographies and to consider the spatial patterning of socio-cultural practices as a guiding principle. Moreover, as stated earlier, in the frame of our research, the lens through which Porto Alegre's elites were defined was that of the "security paradigm" (Goldstein, 2010), for which it was more relevant to track behaviors and ambitions rather than the geographical distribution of income and wealth. For example, the different security systems (walls, cameras, electrified wires, private guards, etc.) employed by wealthy groups to represent themselves by fighting against urban criminality has become a cohesive factor and a common denominator, revealing that these groups "form a cohesive segment in terms of mobilising in defence of their interests" (Cattani, 2011).

Applied methodology

Starting from the assumption that urban elites can be deeply influenced by the perceptual dimension of fear, and that the surrounding built environment is the spatial expression of insecurity, setting up a methodology that combines different tools with that focus was key to understanding urban elites' social distinction and habitus formation (Bourdieu, 1979) as well as their (self)representation, symbolism, lifestyles, and behaviors. Those aspects were suggested by research that highlights how contemporary social transformations are deeply entangled with insecurity, fear, and the resulting imagery, with a strong interrelation with space and the built environment (Caldeira, 2000; Dammert, 2012; Low, 2004). Moreover, along similar lines, a number of studies that specifically address insecurity from a criminological point of view also take into consideration individual, familial, and communitarian features, socio-demographic and personal characteristics, the experience of victimization, social integration, trust in public authorities, and the impact of the media in discussing crime (Dammert & Malone, 2003; Eschholz, 1997; Walklate, 1998).

Such methodology, in our view, should also take note that elites construct, employ, and perpetuate their power and influence, their behavioral mechanisms, social perceptions, and cultural valuation among the larger public (Salverda & Abbink, 2013, p. 3). These factors are deeply connected to the construction of the material, urban-architectural landscape in which the elites live.

Thus, the specific research need to investigate the urban environment and private spaces of wealthy groups, together with the necessity to understand to what extent their behaviors and discourses are influenced by insecurity, led to the construction of a mixed methodology for which a set of tools were borrowed from mainly two disciplines: architecture-urbanism, and anthropology. The investigation called for the application of empirical methods at the crossroads between the two disciplines and aimed at finding effective ways to conduct research on-site so as to uncover the urban geographies of elites. Specifically, the effort of rooting the research methodology in the anthropological approach arises from the concrete need to provide more detailed and contextualized descriptions of the urban elites.

Accordingly, ethnographic fieldwork (along the same lines as Setha Low, 2004–2005) was combined with detailed direct observations: video and photographic surveys with on-site sketching and drawing, structured and extemporary interviews with guided tours, mapping exercises on topographical drawings with perceptual sketching, data analysis with infographics, and geographical spatialization of information with visual and morphological analysis.

The collection of quantitative data was the first step in order to construct an initial definition of urban elites that could be verified through the results of the fieldwork. Institutional sources such as the census (IBGE), government documents, academic research, and available urban plans were the basis for delineating a first profile of the elites and for understanding some of their features (i.e. preferred living locations and average income). From that base, we used two main actions aimed at integrating information about space and about subjective perceptions, experiences, and facts: urban architectural surveys and selected interviews. Spatial analyses were conducted through ad hoc visits to wealthy neighborhoods, after which we selected a sample of dwellings according to the most recurrent typologies (vertical and horizontal gated community, single building and single family house). The goal of this part of the study was to survey the "living space of the elites," how it is constructed and used, its security systems, and the sequence of thresholds, barriers, and control devices that have to be negotiated in order to move from public to private spaces.

In parallel, subjective points of view were collected through a series of video recorded interviews and mapping exercises with selected interviewees. For that purpose, the identification of an appropriate sample was a crucial point. The selection criteria used in order to choose representatives of Porto Alegre's elite were family income, the neighborhood, and the type of dwelling. Once the study areas (wealthy neighborhoods) had been identified, the selection–in line with the criteria described above–was established thanks to a network of local contacts built up before and during the fieldwork, which greatly facilitated access and communication with the interviewees.

In order to obtain an accurate description of the dwelling and to record the interaction with its spaces, video interviews–45 in total over five months by five researchers–were conducted inside homes. Moreover, given the emphasis on the perception of insecurity in the city, three additional exercises were elaborated, aimed at revealing how urban elites behave in their daily life and the relationship between urban private and public spaces.

During the interviews, we asked the interviewee to elaborate two "perceptual drawings" on a blank sheet of paper, which represented the most insecure and the most secure space (Figure 1). Additionally, on a printout map of the area surrounding the dwelling, our interlocutor had to draw a "*dia-a-dia* (daily) map", tracing his or her most recurrent movements in the neighborhood or in the town, and indicating his or her preferred destinations (supermarket, school, bar, gym, etc.). In some instances, when the case study was particularly representative of an emblematic spatial configuration or situation (i.e. the security system in the horizontal *condominio fechado*, or strategies for driving or parking at night) we carried out what we called "safe tours," filming the interviewee while moving around and illustrating his or her defensive daily practices and existing security devices.[3]

Emerging issues

The combination of architectural surveys of private spaces, video interviews, and mapping exercises allowed us to understand some of the relationships between the discourses of urban elites on safety, their perceptions of fear, daily behavior, and the construction of their environment, which led to the identification of different profiles.

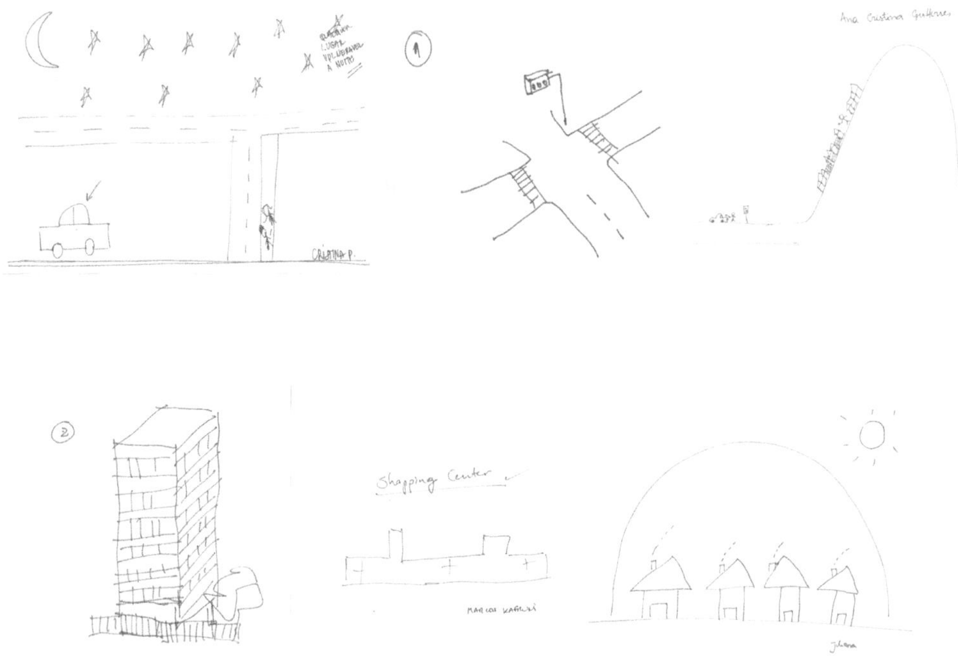

Figure 1. Perceptual drawings.

Examples of "perceptual drawings" of insecure (first row) and secure spaces (second row) made by some of the interviewees representing the urban elites of Porto Alegre. The selection points to some of the most recurrent elements identified by the sample (2014)

Without pretending to be exhaustive, the research output and the application of the above-described methodology provided some interesting results.

The *condominio fechado* is the widely preferred housing type among Porto Alegre's urban elite; both because it represents a clear status symbol and for security reasons. As Caldeira (2000) points out, the diffuse and massive construction of this type of model in most Brazilian cities since the 1980s stems from a reaction to crime and the lack of control by public authorities (police) regarding safety. This type of gated building is a private separate domain that contains equipment and facilities normally found in the city. Gardens, leisure spaces–such as a *salão das festas* (party room) or *brinquedoteca* (play room)–and sports facilities can be found inside the secured walls of the *condominios fechados* and their presence, richness, number, and complexity are direct indicators of status. In parallel, the presence, evidence, and sophistication of security systems that surround the *condominio* are also seen as strong signs of wealth (Facchetti, 2015).

Even if this way of living is not preferred by the super-rich, who tend to live in segregated villas outside the city, it continues to be a successful model for the upper middle class. The *condominio* offers an aggregated way of living that guarantees a high class "affordable" status, safety, and in certain cases the access to a community of pairs.

For example, Juliana and her neighbors, who live in a horizontal gated community on the east side of Porto Alegre, remarked that inside the *condominio fechado* they feel safe, thanks to the "scary" defensive system around it. This allows them to avoid any security measures at home–"we leave windows and doors open" (Juliana, 2014, interview)–which favors the creation of a sense of community among the inhabitants of the compound who have a "very cooperative life" (Ana Cristina, 2014, interview).

Despite the early individualistic tendency described by Caldeira (2000) and the problems deriving from such an enclosed model, the research in Porto Alegre shows a positive trend toward building a "sense of community" inside the *condominios*, where elites can live among people belonging to the same group, sharing a similar lifestyle, and (in some cases) establishing a feeling of solidarity between each other. At the same time, the creation of such an enclosed world, defined as "the bubble" (Juliana, 2014), emphasizes the disconnection from the outside, which is generally perceived as dangerous: "the entire city at night is dangerous" (Cristina, 2014). "We spend our life here, our world is here," claims Cristina. For that reason, together with the distrust of public control and the general rise in rates of violent crime, most of the gated communities we studied had plans to reinforce the external defensive systems located along the perimeter of their properties; increasing the number of cameras, fences, and control devices.

That trend is confirmed by the real estate market and experts in the field (real estate agents, security system seller; Katu, Julio Cesar, 2014, interview) who highlighted how new gated communities tend to incorporate as many facilities as possible in order to offer an all-in-one, 24/7, introverted world, where not only living and leisure spaces are provided within the boundaries of the *condominio*, but also offices and shops (i.e. Barra Shopping), offering an alternative safe enclosed environment as opposed to the city.[4] Observing the single apartment building type of dwelling,[5] the applied methodology provided some other interesting results. On the one hand, representatives of the urban elites that live in wealthy, socially homogeneous neighborhoods (Rica and Simone, 2014, interview) such as Tres Figueiras, Bela Vista, and Higienopolis, reported a greater sense of insecurity, even at home; the safe space is restricted to the apartment or the private room. They are obliged to use a greater number of outside facilities, since apartment buildings cannot offer as many as *condominios* can. Feeling more exposed to violent threats, those interviewees tended to develop elaborate, personal behavioral strategies against crime (how to drive, park, enter their home) and to adopt security measures at home (door spy-hole, cameras, weapons). "If he [the criminal] enters from that door, I can easily shoot" said Simone, who keeps a gun in the drawer of the bedside table, as she drew in representing her bedroom as the most secure space. At the same time, due to restricted shared spaces, inhabitants of apartment buildings seem not to develop a communitarian way of living with their neighbors, but instead to enact their social life in a limited number of places outside the building, often in the neighborhood: "my entire life is within a very limited perimeter" (Simone, 2014, interview). Spatial surveys show the correspondence between the greater feeling of insecurity and the architectural-urban form.[6] Vertical typologies provide fewer, smaller, more fragmented and enclosed communal spaces with lower mutual visibility when compared with horizontal gated communities.[7] Moreover, due to the limited number of inhabitants, the security system is relatively simple (Single fence with barbed or electrified wire, double entrance gates, CCTV system (around 10 cameras) and one doorman) which increases its potential vulnerability.

On the other hand, representatives of urban elites living in apartment buildings in mixed neighborhoods, such as the city center or next to a *favela*, present different stories. Their security system, both at the building and at the apartment level, is limited or absent (single gate, no cameras). They do not develop particular security strategies and they travel by car, public transport, bike, or on foot in a wide range of socially mixed areas of the city. Their daily contact with different people in shared or public spaces makes them aware of the vicious circle created by social separation and visually aggressive security systems, which paradoxically generate a greater sense of insecurity (Daniel, Rodrigo, and Fiapo, 2014, interview). Consequently, contrary to the actual trend, their discourse on urban development calls for the creation of spaces for social interaction, the need to keep public spaces permeable, and the public control of gated community development projects. In their opinion, social mix, interaction, and control comprise effective ways to prevent crime, something confirmed by various experts.

Opposing that vision, there are urban elites who have either directly or indirectly experienced violent crime and who live "among peers" in exclusive, separated environments. According to the sample we interviewed, the gated method of living is seen as a "necessary evil," as explained by Juliana (2014). While understanding the reasons behind the high crime rates in Brazil, they seemed to justify racial[8] or general distrust of "the other" based on the assumption that criminals are everywhere and "you have to pay attention anytime" (Rica, 2014). "Here you cannot recognize a mugger. He can be dressed like you, like me, like a normal person. By the way, the most dangerous ones are the well-dressed. You cannot look at a mugger's face and guess he is a mugger," claims Simone (2014). Consequently, the necessary, living, urban model can only be the private, gated, and highly secured sequence of spaces where the "community" can live, work, shop, and spend free time, separated and detached from low-income neighborhoods and the city. The proliferation and success of that model, despite the increasingly poor living standards offered by real estate construction companies (i.e. Goldstein and Cirela), is in fact a sign of a trend that is taking root where the lack of public authority is compensated for by an aggressive security real estate market supported by the media.

However, it is important to point out an important behavioral difference among our interviewees (which needs to be supported by future research on the basis of a wider sample) that suggests different generational attitudes toward the issue of security. According to our sample, whereas individuals around and above the age of 60–who were adults during the dictatorship in Brazil–showed little worry on a personal level, by contrast those between 35 and 60 felt constantly threatened, which had a major impact on their daily life (Hilda, Potiguara, Gabriela, and Rica, 2014, interview). That difference not only influences lifestyles and individual choices, but also discourses about safety, about "the other," and about causes of and remedies for dealing with the issues.

Security systems play a key role in understanding socio-spatial relations and the logic behind the formal-informal construction of the city. Our spatial surveys show that in fact preventive security concentrates on and reveals a form of "climax" at the boundary (the fence) between private and public spaces; however, it is a complex system that should be observed on multiple scales and involves cameras, security devices, lighting, doormen, guards, and the patrolling of paths (Figure 2).

Interviewees point out that the security "show off" concentrated on the boundary is necessary to prevent common crime, but is ineffective against organized crime. For that reason, multiple strategies should be displayed from the dwelling to the urban realm. Security systems are necessary "palliatives" to combat criminality, even if there is clear consciousness of the fact that they act as reminders

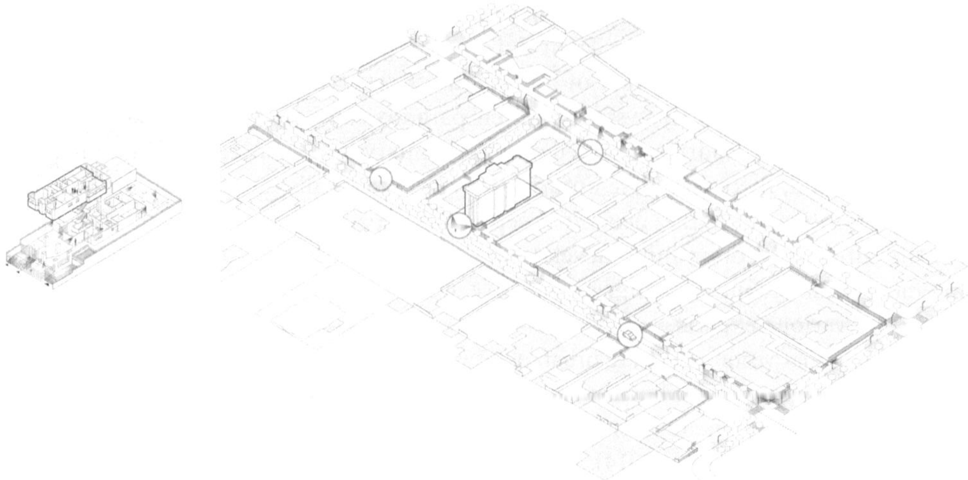

Figure 2. Security system.
Axonometry highlighting the defensive system of one of the case-study dwellings in Porto Alegre at the building and at the neighborhood level (2017)

of a latent but constantly present risk, generating a "vicious circle in which the increased precautions feed back on insecurity and in turn generate the need for increased protection" (Dammert, 2012, p. 19).

At the same time, an (apparently) contradictory tendency is the "beautification" of the fence. Referring to a couple walking on the streets of Porto Alegre, Fiapo mentioned: "they were commenting along the road, how it was getting pretty with all the same type of fences … we laughed seeing how the thing was turning out" (Fiapo, 2014, interview). While on the one hand the security apparatus is meant to be visible and frightening, on the other hand we observed the tendency to use transparent armored glass instead of walls or iron bars, or to use plants and other ornaments to "beautify" the surveillance apparatus. In many analyzed cases, the installation and use of devices, as well as the use of weapons, is also at the edge of legality, revealing the existence of a high degree of informality that characterizes crime prevention.

In this regard, it is relevant to highlight a counter-case study, such as the one of a wealthy family who decided to build an apartment building without security systems next to a small *favela*. The family theorized a principle of *boa vizinhança* (good neighborhood), which involves establishing daily relationships with the people living in the *favela* as a precise strategy for guaranteeing their safety. For that reason, any security system around the building "would be seen as aggression" toward the neighbors (Potiguara, 2014, interview).

Conclusions

The different profiles of Porto Alegre's urban elite that emerge from the research "Reducing Boundaries" highlight a variety of elements that are deeply entangled with the specific socio-spatial environment in which our sample live.

The investigation of living spaces for the wealthy shows that their form is influenced by issues of safety, and that the definition of Porto Alegre's urban elite cannot be dissociated from that. Therefore, on a large scale, the construction of the city can only be seen as the result of such dynamics and it cannot be understood without a contextual approach. Spatial surveys, interviews, behavioral maps, and safety tours were key to understanding the complex system of relations composed of both objects and behavioral patterns.

In that respect, the combination of spatial observations and an ethnographic approach was fundamental to obtaining an understanding of the relationship between specific, measurable socio-spatial, features (data and spatial analysis), the perception of fear and the consequent re-elaboration (semantically and symbolically), and the re-shaping (materially) of space (Casagrande, 2016). That relationship comprises the comprehension of personal and collective experiences, the imagined horizons and futures, and the understanding of "the other."

We can highlight a number of potential areas for studying urban elites using empirical applied methods and the lens of fear and insecurity.

Fieldwork based on the combination of interviews and spatial readings proved to be effective in reaching a complex definition of Porto Alegre's urban elite while considering the behavioral patterns and discourses that depend on family relations, educational trajectories, political structures, lifestyles, individual and group perceptions, and interpretations. Since those elements cannot be removed from a definition of elites based only on data and economic indicators, the described methodology was fundamental for completing, enriching, and correcting the profile delineation. Moreover, the combination of the two approaches, derived from anthropology and architecture-urbanism, has been shown to be fruitful and complementary. While the first allowed a deeper understanding of the social dynamics, the second revealed their strong relationship with spatial configurations and constructions.

Another potentially strong point of the study is the specific angle offered by the theme of insecurity, which highlights the need to understand elites, their behavior, and reproduction, beyond the usual economic parameters. In fact, the use of different thematic lenses can help in reading and interpreting the positioning of the elites within a certain society, their self-perception, and definition. Moreover, since fear and security by definition specifically operate on a perceptual level, these topics draw attention to the influence of elements that go beyond rational and material criteria. Thus, paying particular attention to socio-cultural and perceptual aspects, socio-spatial strategies, and the practices of urban elites–as well as looking at non-elite groups who aim to emulate elite behavior in order to reach similar positions–contributes to the understanding of collateral aspects that may play a key role in achieving a precise delineation of elites.

The research findings described above allowed us to make a distinction between elites living in different urban conditions in Porto Alegre, and to acknowledge that within the same urban elite (area/group) there are sub-groups that do not necessarily share the same interests or way of life. The self-definition and social status of such sub-groups vary, and intra-elite competition and lifestyle emulation are relevant dynamics that need to be considered. Lastly, the results suggest that elites still largely rely on direct or meta dominance over the lower population strata.

These differences are therefore important, as they portray a complex picture in which urban elites should be understood in their background conditions. The effort of drawing transversal lines should help in achieving a multifaceted definition that can offer a precise profile and at the same time substantial information for comparative studies. The need for additional investigations, where the explained methodology might be further applied, tested, and improved, could also contribute to a better understanding of the complexity and the key position of elites as a driving force of urban change.

Notes

1. The comparative research focussed mainly on Porto Alegre (Brazil), comparing the city with two other case studies. Brussels (Belgium) and the Veneto Region (Italy). The three leading institutions involved in the project are: IUAV University of Venice (IT), Latitude Platform (BE) and Lasalle University (BR). The research has received funding from the People Programme (Marie Curie Actions) of the European Union's Seventh Framework Program FP7/2007–2013/under REA grant Agreement n. 612,063. The research results can be found in the book "Reducing Boundaries: Understanding Exclusion Through Security Defensive Systems in Wealthy Urban Areas" (2017).
2. In Porto Alegre (1,409,000 inhabitants in 2010), from 2004 to 2014, there were 233,912 robberies (7.4% in high-class neighborhoods), 69,477 car robberies (7.4% in high-class neighborhoods), 316 larcenies involving fatalities (11.1% in high-class neighborhoods), and 4,345 fraudulent homicides (3.4% in high-class neighborhoods). For almost half of the recorded incidents, localization data is missing (data: Policia Civil of Porto Alegre). Moreover, since 2015 violence has increased in Porto Alegre, being declared the 36th most violent city in the world (Conselho Cidadão pela Seguridade Social Publica e Justiça Penal, Mexico).
3. A "safe tour" differs from a "guided tour," because it is meant to specifically show the defensive practices and behaviors that the interviewee adopts in order to feel safe. While guided tours are illustrative–the interviewee describing what he or she sees as external factors–safe tours illustrate daily practices on a "live" basis.
4. Recent real estate projects show a tendency for preferring mixed, gated communities where different functions are combined (shops, offices, and apartments) and where security systems tend to be invisible rather than shown off.
5. The single apartment building type is characterized by the limited outer private space around it. In fact, most of them, especially in central neighborhoods, were built on single family house plots.
6. Spatial surveys are meant as the combination of architectural on-site sketches and measurements, photographs, and video: media that allow a detailed reconstruction of the spatial conditions of a place as well as the identification of physical aspects that suggest specific practices (i.e. the use of walls, cameras, etc. as a defensive approach and as a reaction to insecurity).
7. This is also evident when observing internal movements that do not provide notable opportunities for interaction: entering underground parking by car, going to the elevator, arriving at the landing with only a few apartments' front doors. Despite the larger number of shared spaces, the phenomenon of social fragmentation can also often be observed in vertical gated communities (i.e. Gabriela: interview, 2014).
8. Since according to statistics most of the criminals are black or mulatto, that fact is assumed as a reason for distrust: "It was a black guy. You can say I have preconceptions, but the truth is that today I have preconceptions because both times I was mugged it was by poor people," Fabio: interview, 2014.

Disclosure statement

No potential conflict of interest was reported by the author.

References

Bauman, Zigmund. (2001). *Community. seeking safety in an insecure world*. Cambridge: Polity Press.

Bauman, Zigmund. (2003). *City of fears, city of hopes*. London: Goldsmiths College, University of London.

Beck, Ulrich. (1992). *Risk society: Towards a new modernity*. London: SAGE.

Bourdieu, Pierre. (1979). *La distinction: A critique sociale du jugement*. Paris: Les Editions de Minuit.

Caldeira, Teresa. (2000). *City of walls. crime, segregation and citizenship in São Paolo*. Los Angeles: University of California Press.

Casagrande, Olivia (2016). Fear Landscapes. Working paper.

Casati, Roberto. (2012). Urban safety and security. In SUPSI – Laboratory of Visual Culture (Ed.), *Mobile A2K methodology guide- mobile access to knowledge, culture and safety in Africa* (pp. 12–14). Lugano: SUPSI.

Cattani, Augusto. (2011). *The old class: The dark side of the rich Latin American Centre*. Oxford: University of Oxford.

Coloquhoun, Ian. (2012). *Designing out crime. Creating safe and sustainable communities*. Princeton: Architectural Press.

Dammert, Lucia. (2012). *Fear and crime in Latin AmericaRedefining state-society relations*. London & New York: Routledge.

Dammert, Lucia, & Malone, Mary Fran. (2003). Fear of crime or fear of life? Public insecurities in Chile. *Bulletin of Latin American Research*, *22*(1), 79–101.

Eschholz, Sarah. (1997). The media and fear of crime: A survey of the research. *Journal of Law and Public Policy*, *9*(1), 37–59.

Facchetti, Andrea. (2015). Reading security. Imaginaries of exclusivity and exclusion in Porto Alegre. In Emanuela Bonini Lessing (Ed.), Urban safety and security (pp. 37–48). Milan: Franco Angeli.

Giddens, Anthony. (1999). *Runaway world: How globalization is reshaping our lives*. London: Profile.

Goldstein, Daniel. (2010). Toward a Critical Anthropology of Security. *Current Anthropology*, *51* (4), 487–517.

Kesteloot, Christian. (2013). Socio-spatial fragmentations and governance. In Eric Corijn & Jessica van de Ven (Eds.), *The brussels reader. A small world city to become the capital of Europe* (pp. 110–149). Brussels: VUB Press.

Low, Setha. (2001). The edge and the center: Gated communities and the discourse of urban fear. *American Anthropology*, *103*(1), 45–58.

Low, Steha. (2004). *Behind the gates. Life, security and the pursuit of happiness in fortress America*. New York: Routledge.

Lupton, Deborah (Ed.). (1999). *Risk and socio-cultural theory. New directions and perspectives*. Cambridge: Cambridge University Press.

Lupton, Deborah. (2013). *Risk* (2nd ed.). London & New York, NY: Routledge.

Newman, Oscar. (1972). *defensible space: Crime prevention through urban design*. London: MacMillan.

Povinelli, Elizabeth. (2011). *Economies of abandonment: Social belonging and endurance in late liberalism*. Durham: Duke University Press.

Reid, Lesley Williams. (2003). *Crime in the city: A political and economic analysis of urban crime*. El Paso: LFB Scholarly Pub.

Reiner, Robert. (2007). *Law and order: An honest citizen's guide to crime and control.* Cambridge: Polity Press.

Salverda, Tijo, & Abbink, Jon. (2013). *The anthropology of elites: Power, culture, and the complexities of distinction.* Basingstoke: Palgrave Macmillan.

Vanin, Fabio. (2015). Physical and ephemeral devices for urban security: The case of Luanda. In Emanuela Bonini Lessing (Ed.), *Urban safety and security*(pp. 81–92). Milan: Franco Angeli.

Vanin, Fabio. (2015). Questioning the Urban Form: Maputo and Luanda. In Carlos Nunes Silva (Ed.), *Urban planning in lusophone African countries*(pp. 161–170). London: Ashgate Publishing Ltd.

Wacquant, Loïc. (2006). *Parias urbains. Ghetto, banlieues, état.* Paris: La Découvert.

Walklate, Sandra. (1998). Crime and community: Fear or trust? *The British Journal of Sociology, 49*(4), 550–569.

Reframing labour market mobility in global finance: Chinese elites in London's financial district

Sarah Hall

ABSTRACT

In this paper, I use the case of elite Chinese financial mobility to London's financial district to argue that comparatively neglected forms of elite financial migration from beyond the Global North provide important insights into the changing geographical form, and labour market practices within, leading international financial centres. Empirically, the paper identifies a distinctive geographical footprint of Chinese financial mobility to London in terms of both financial services activity and residential choices. Theoretically, I argue that explaining this geography requires work on highly skilled migration that typically privileges the economic motivations underpinning mobility to engage more fully with wider debates in economic geography that examine the inter-dependencies and inter-relationships between states and markets. These findings raise important questions surrounding the durability of Chinese finance in London, its relationship to global finance in London more generally, and wider understandings of elite financial labour markets.

Introduction

The international mobility of financial elites has attracted growing academic, popular and political attention in recent years (see, for example, The Financial Times, 2016). Research in this area emphasises how such mobilities shape the flows of knowledge and expertise through which international financial centres (IFCs) are reproduced (Beaverstock, 1996; Beaverstock & Hall, 2012; Sassen, 2000; Thrift, 1994). This work builds on, and contributes to, a wider literature on elite, or what Kunz (2016) terms "privileged mobilities" (see also Sklair, 2001; Smith, 2001). However, the geographical focus of this work remains limited. Studies typically focus on flows of migrants either between IFCs in Europe and North America (Beaverstock, 2005; Ryan & Mulholland, 2014) or on the mobility of individuals from the Global North to rapidly growing economies, often in Asia and the Middle East (Beaverstock, 2002). As a result, little is known about what happens to theoretical understandings of the rationale behind, and implications of, international mobility when different geographies of mobility are in play (although for exceptions see Knowles, 2017; Moore, 2016).

In response, and inspired by calls to take the urban dimensions of mobility seriously (Smith, 2001), this paper reports on original research conducted into flows of Chinese financiers from mainland China to London. This empirical focus is significant for two reasons. First, figures show that Chinese born nationals now make up the largest source of migrants to the UK (Dominiczak, 2013) and yet understandings of Chinese migration to the UK remain largely rooted in earlier waves of immigration that were characterised by migrants working in lower paid service sector jobs, particularly the restaurant and food trade (Knowles, 2017; Pieke, 2007). Second, demand for elite Chinese financial labour in London has grown significantly since 2012, following London's development as a leading western centre for the internationalisation of China's currency, the renminbi (RMB) (Hall, 2017a, 2017b; Töpfer & Hall, 2017).

By reporting on findings relating to the motivations, experiences and future planned trajectories of Chinese financiers in London, two main findings emerge. First, Chinese financial migration into London has a distinctive geographical footprint in terms of both financial services activity and the residential choices of Chinese. This points to the emerging, ways in which Chinese finance is beginning to shape London's wider economy. This is not limited to the financial services activity at the heart of RMB internationalisation but, also extends to Chinese investment in London real estate and the significance of Chinese retail spending in the development of luxury consumption practices in London (on which see Atkinson et al., 2017; JLL, 2015; respectively).

Second, the rationale behind elite Chinese financial mobility in London does not sit neatly within existing understandings of elite mobility and expatriation that emphasise the economic rationales underpinning such mobility, particularly at the firm level (see Beaverstock, 2004 for example). Whilst my analysis shows that for some financiers, their motivations for working in London were underpinned by economic motives that dominate existing theoretical explanations of elite migration, political motives also need to be more fully theorised than has been hitherto the case. In this sense, the work and mobility of Chinese financial elites was closely tied into the Chinese state project aimed at having strategically important IFCs such as London tied into the RMB internationalisation process (see also Töpfer & Hall, 2017). This finding suggests that work on elite migration could usefully draw on wider debates in economic geography that seek to understand the inter-dependencies and inter-relationship between states and markets (see O'Neill, 2008 for example) as the Chinese and UK state facilitate Chinese elite migration to London and in turn this migration supports wider state policies relating to both RMB internationalisation and the wider competitiveness of London as an IFC. Taken together, these findings raise important questions concerning the future trajectory of Chinese elite migration to London as well as broader points concerning London's future development as an IFC following the referendum vote to leave the EU by the UK in June 2016 (Hall and Wójick, 2018).

I develop these arguments over four sections. Next I locate the mobility and expatriation of Chinese financiers within the wider literatures on elite migration and expatriation in general, and Chinese mobility into London in particular. I then reveal the geographies of Chinese finance and financiers in London, revealing how Chinese financial elites, and the financial institutions they work for, are concentrated in the historic centre of London's financial district and inner London boroughs with important consequences for the relationship between Chinese financial elites and elite labour markets more generally in London. In the next substantive section, I explore the different routes by which these geographies have been created through Chinese elite migration and expatriation into London. Here, I examine how Chinese financial elites in London are typically at an earlier career stage than elites from Europe and North America that have been more widely studied. This career stage reveals close links between educational background and elite Chinese mobility in ways that trouble assumptions of a relatively seamless transition from selected higher education institutions and elite financial labour markets. I conclude by reflecting on the implications of these findings for understandings of labour market mobility in global finance in general, and the reproduction of London as an international financial centre in particular.

Mobility and international financial centres: the case of elite Chinese urban economies

My analysis of the mobility of Chinese financiers into London takes as its starting point the growing academic interest in highly skilled labour between nation states over the least twenty-five years (see for example Iredale, 2001; Koser & Salt, 1997). Much of the early research in this area focused on the mobility of individuals with higher educational qualifications within the transnational corporations they worked for (Findlay, Li, Jowett, & Skeldon, 1996). The geographical focus of this work on mobility was dominated by migration from and between global cities within advanced economies (Sassen, 2000). For example, work on inter-company transferees, particularly in its earliest phases, emphasised the ways in which individuals from Europe and North America especially were sent by their employers as expatriates to newly opened offices in rapidly growing economies in order to ensure that similar standards of service were maintained in newly opened offices as compared to the firm's headquarters (Beaverstock, 2004).

Building on this early work, research increasingly began to interrogate who counted as a highly skilled migrant by studying more closely the types of assignments being undertaken by migrants, and the geographies associated with these (Findlay & Cranston, 2015). Reflecting this questioning of the term "highly skilled", as well as a desire to acknowledge the privileged economic and political position of the Chinese nationals involved in this research, I use the term elite mobility rather than highly skilled migrant in my analysis (see Savage & Williams, 2008 for a wider discussion of the importance of studying elites within the contemporary global economy). Research also began to move beyond mobility driven through inter-firm transfers to examine the mobility of the self-employed and entrepreneurs (Ley, 2010; Saxenian, 2006). This change in focus brought with it a greater attention on the rationale for mobility moving beyond a focus on disseminating best practice from the headquarters of transnational corporations typically through flows of migrants from the Global North. Rather, research became increasingly concerned with understanding the ways in which migration can be used to foster economic development and innovation within mature economies (Faulconbridge, Beaverstock, Hall, & Hewitson, 2009), often raising concerns about a potential loss of "talent" in home economies (Florida, 2007; Saxenian, 2006).

However, in the case of financial services, the geographical imaginations of work on elite migration have not widely embraced the more heterodox geographies of this work on migration and expatriation from beyond the Global North. Typically, research has focused on the heartlands of global finance through research into the nature of financial centres in Western Europe and North America, or on the circulation of elites from these leading financial centres to other spaces of global finance through processes such as expatriation (Beaverstock, 2002, 2005). This builds on a wider understanding in academic and policy circles that has frequently identified a ready supply of graduate labour as a key factor in explaining the competitiveness of leading financial centres (Beaverstock & Hall, 2012; Z/Yen, 2016). However, as Robinson (2002, 2005)) has argued, much of this work is characterised by a distinctive geographical focus on the mobility of elites from the Global North to other parts of the world. As a result, much less attention has been paid to financial elites beyond Western Europe and North America, their mobility patterns and working practices (although see Bassens, Derudder, & Witlox, 2011, 2010; Lai, 2012; Krijnen, Bassens, & Van Meeteren, 2017; Kleibert & Kippers, 2016).

Reframing elite mobility between financial centres

The case of Chinese migration into London, therefore, raises important questions about how well our existing understandings of the motivations, experiences and trajectories of migration within the financial services sector can explain elite Chinese mobility into London and its role in IFC formation and reproduction. Addressing these concerns is also important in terms of updating understandings of Chinese migration to London, and the UK more generally, that remain largely rooted in more historic migration flows of Chinese nationals working in lower paid service and hospitality sectors. In this respect, it is important to locate the forms of mobility studied in this paper within the wider transformation of Chinese migration from a process that was only peritted from Hong Kong and other overseas Chinese territories, to one open to a far wider segment of the Chinese population, particularly the rapidly growing Chinese middle class and young people undertaking mobility for studying (Pieke, 2007).

After the Second World War, the People's Republic of China (PRC) effectively banned all official emigration until the 1960s (Pieke, 2007). As a result, there were only modest international migration flows from Hong Kong and Taiwan to the UK, reflecting their unique political status as overseas territories. In contrast to the Chinese migrants who form the basis of this paper, in terms of occupation, this migration stream to the UK was characterised by individuals working in relatively low paid, service sector jobs, particularly in the catering sector. This pattern of migration changed from the 1980s onwards as the PRC began to relax the regulations surrounding migration. Following this, migration became an option for Chinese citizens from mainland China who could demonstrate they could legitimately enter a foreign country (Xiang, 2003). As a result, two migration flows from mainland China to the UK can be identified from the 1980s onwards as part of what Pieke (2007) terms the "new Chinese migration order."

First, Chinese elite migration grew significantly from the 1990s onwards, reflecting the wider opening up of the Chinese economy associated with domestic economic reform. As a result, an increasingly entrepreneurial elite Chinese middle class emerged in China, for whom international migration represented a significant economic and cultural opportunity (Tomba, 2004). Work on the resulting diaspora of skilled Chinese migrants has cautioned against simply inserting this group of migrants into existing frameworks for understanding elite migration. For example, Saxenian's (2006) seminal analysis of the role of migration from China (amongst other countries) in the labour markets of Silicon Valley recasts this migratory flow away from understandings of an engineering brain drain from China and instead develops understandings of brain circulation as Chinese migrants return to China, and in so doing shape economic growth in China and Silicon Valley (see also Ong, 1999).

Second, building on a more established history of migration from the 1960s onwards from Hong Kong to the UK for educational purposes, significant research attention has been paid to the growth of education related migration from mainland China from the 1990s onwards (Waters, 2006). Again, political economic changes within China are important for understanding the growth and transformation of this form of migration. The rise of a Chinese middle class alongside Chinese entrepreneurs has increased the number of Chinese households who are able to afford foreign education for their children. This has expanded beyond a preference for postgraduate education, largely in North America in the 1990s (Pieke, 2007) to include undergraduate degrees, compulsory schooling and short-term English language courses in a more diverse range of geographical locations, particularly in the UK and western Europe (Bohm et al., 2004).

Taken together, this work is important in revealing the particular political, economic and cultural contexts shaping contemporary forms of elite Chinese migration. However, less attention has been paid to the migration of Chinese born nationals into financial and related professional services, and it is this oversight that I address in the rest of this paper.

Methods

The paper draws on two main strands of research. The first seeks to quantify and document the spatiality of elite Chinese financial migration in London. This seemingly straightforward task becomes more challenging given the acknowledged difficulty of obtaining accurate data on elite migration to the UK more generally, particularly in terms of the nationality of migrants (Findlay & Gould, 1989). As a result, it is difficult to use official data sources to disaggregate the number of Chinese migrants working in London by sector. For example, because the UK census is only conducted every 10 years with the last one being held in 2011, the recent impact of RMB internationalisation on Chinese migration to the UK from 2012 onwards is not picked up in census data. In order to overcome the temporal limitations of other forms of migration data, my research draws on an analysis of new national insurance registrations by adult overseas nationals entering the UK by year of registration. This data is provided by the Department for Work and Pensions and has the advantage of being updated annually. A National Insurance (NI) number is needed to pay tax in the UK and hence captures economically active migrants and can be located geographically within London. However, it is not disaggregated by sector so inferences have to be made based on the location of migrants in order to estimate the sectors in which they work.

Therefore, in order to better understand the motivations behind the migration patterns shown in quantitative data, the second methodology used in my research are semi-structured interviews with Chinese migrants working in London's financial district. This method also posed important methodological issues including securing access and the situated nature of the knowledge produced, both of which have been well documented in relation to elite corporate interviews (see Schoenberger, 1991, 1992; and McDowell, 1992; in general and, 1998; on the case of financial elites in particular). I used several strategies to access individuals in order to undertake thirty interviews between September 2014 and June 2015 with Chinese financiers working in a range of RMB denominated financial markets in London. Conventional ways of accessing financial elites do not work in the case of Chinese migrants working in financial services in London. For example, typically the more senior employees within financial organisations can be accessed via their firm's corporate website. However, the overwhelming majority of the migrants I interviewed were at an early career stage (twenty five out of thirty) and hence did not appear on lists of companies managing boards. Meanwhile, the websites of Chinese financial institutions operating in London's financial district, particularly the state owned commercial banks, remain rudimentary in nature and do not contain staff names and contact details.

In response, I used the business orientated social network LinkedIn.com as my primary way of accessing initial research participants. Through my own LinkedIn account, I searched for individuals working in the state owned commercial Chinese banks operating in London, setting out the details of my research and asking if they would be prepared to meet for an interview. Given the unsolicited nature of my enquiry, I was expecting a very low success rate. However, to my surprise this proved very productive and I secured interviews with individuals working at different levels within these banks. During the interview, I asked why they had been keen to participate. The value placed on the cultural capital associated with education amongst Chinese migrants (Waters, 2005) and my own position as a university professor at a large UK higher education institution with operations in China emerged as key factors in them agreeing to the interview (this echoes McDowell, 1998 discussion of how her own career trajectory facilitated research entry into London in the late 1990s). It also became apparent that as a female researcher, my interviewees positioned me as an outsider, as they themselves felt, within London's financial district. So, whilst, they understood my position as being different from the dominant masculine subjectivities of the City, they felt they shared with me a sense of being "other" through their position within a relatively small elite migrant group, and this was frequently used in their responses to my questions. This experience builds on the wider feminist literature on the intersectional positionality of interviewers, and female interviewers in particular as legitimacy was conferred on me by my respondents because of my job in education whilst simultaneously being seen as non-threatening and relatively powerless (see Pratt, 1993; England, 1994). Comments such as the following were typical in this respect

> "You'll understand the difficulties of getting on in a labour market like this when you don't quite fit in – I don't always feel like employers know how to handle my difference – different background, different skills and different aspirations"
>
> (Associate, European investment bank, London June 2015)

From these initial contacts, I used a process of snowballing to access other Chinese migrants working in a range of financial institutions that were both Chinese and internationally owned. In so doing, I was made aware of several online and "real world" social networks for Chinese financiers in London that were either education based (e.g. alumni groups from Masters of Finance degrees at the London Business School) or aimed at supporting Chinese migrants in their career development (e.g. The Association of Chinese Financial Professionals in the UK). These organizations were important in terms of recruiting research participants and reflected the demography of Chinese born financiers working in London with most interview respondents aged under thirty-five. Interviews lasted between thirty minutes and two and a half hours, were conducted in English and were recorded and transcribed for subsequent analysis within a grounded theory framework. Details of my interviewees are provided in Table 1.

Placing elite Chinese financial migration in London

Figures from the Office for National Statistics point to a small but rapid growth of Chinese migration to the UK, and Chinese migrants working in financial services in London in particular. For example, the proportion of Chinese migrants working in all sectors across the UK remains small, estimated officially to be 0.72% of the population of England (379,503). Within this, London has the largest number of Chinese migrants (124,250 or 1.52% of the population, although estimates of undocumented Chinese migrants suggest the figure is closer to 300,000 (Knowles, 2015). Indeed, in 2014, more migrants arrived in the UK from China than from any other country (more than 40,000), and yet they remain one of the least studied migrant groups (Knowles, 2015). Further corroborating this upward trend in Chinese migration to the UK, reports from the executive search sector point to the significant increase in the number of mandarin speakers working in and demanded by London's financial services sector (Mandarin here being significant as it has become increasingly important relative to Cantonese that typified earlier waves of Chinese migration from Hong Kong) (Astbury Marsden, 2014).

New national insurance registrations allow us to be more specific in terms of charting the growth of Chinese migrants working in London from 2011–12 onwards increasing from 2830 new NI registrations in 2011–12 to 4404 registrations in 2015–16, an increase of twenty-two per cent (see Figure 1). The timing of this increase fits with London's growing role in RMB internationalisation from 2011 onwards and it is therefore likely that employment in financial and related professional services accounts for at least part of this increase. There is also a geography to NI registrations of Chinese migrants working in London. As Figure 1 shows, the boroughs showing the largest percentage increase in annual NI registrations for Chinese migrants between 2002–3 and 2015–16 are Camden (438%), Westminster (406%), Kensington and Chelsea (332%), Hammersmith and Fulham (246%) and the City of London (220%). These boroughs are typified by their clustering of young urban professionals (Thrift & Leyshon, 1992), echoing the age demographic of elite Chinese financiers working in London as well as the global super rich (Atkinson et al., 2017). The significant increases in the City of London and Tower Hamlets (126%) are particularly important for the purposes of the arguments in this paper since these are the boroughs associated with

Table 1. Educational and career history of research participants (some details have been redacted to ensure the anonymity of individuals. Source: Author's research).

Job title	Educational background	Current employer
Associate	Undergraduate degree, China Postgraduate degree, UK	European investment bank
Trader	Undergraduate degree China	Chinese bank
Associate	Undergraduate degree China, post graduate degree UK	European investment bank
Associate	Undergraduate degree UK	European asset management firm
Analyst	Undergraduate degree China, postgraduate degree UK	European investment bank
Analyst	Undergraduate degree China, postgraduate degree UK	UK bank
Analyst	Under and post graduate degrees in China	Chinese bank
Analyst	Undergraduate degree China, postgraduate degree UK	UK bank
Analyst	Undergraduate degree China	International finance organisation
Analyst	Undergraduate and postgraduate degrees in China	International finance organisation
Analyst	Undergraduate degree China	European investment bank
Analyst	Undergraduate degree China	European investment bank
Analyst	Undergraduate degree China	Chinese bank
Analyst	Undergraduate and post graduate degree	Chinese bank
Analyst	Undergraduate degree China	Chinese bank
Associate	Undergraduate degree Hong Kong	UK bank
Associate	Under and post graduate degrees UK	UK bank
Associate	Undergraduate degree China, post graduate degree UK	UK bank
Analyst	Undergraduate degree China	UK bank
Analyst	Undergraduate degree UK	European investment bank
Associate	Undergraduate degree UK	European investment bank
Associate	Undergraduate degree China	Organisation to support Chinese finance professionals in London
Associate	Undergraduate degree UK	Organisation to support Chinese finance in London
Analyst	Undergraduate degree China	Chinese bank
Associate	Under and post graduate degrees in the UK	European investment bank
Analyst	Undergraduate degree China, postgraduate degree UK	Boutique investment firm
Associate	Undergraduate degree China, postgraduate degree UK	Boutique investment firm
Corporate lawyer	Undergraduate degree China, postgraduate degree UK	International law firm
Corporate lawyer, China specialist	Undergraduate degree Hong Kong, postgraduate degree UK	International law firm
Capital markets lawyer	Undergraduate degree UK	International law firm

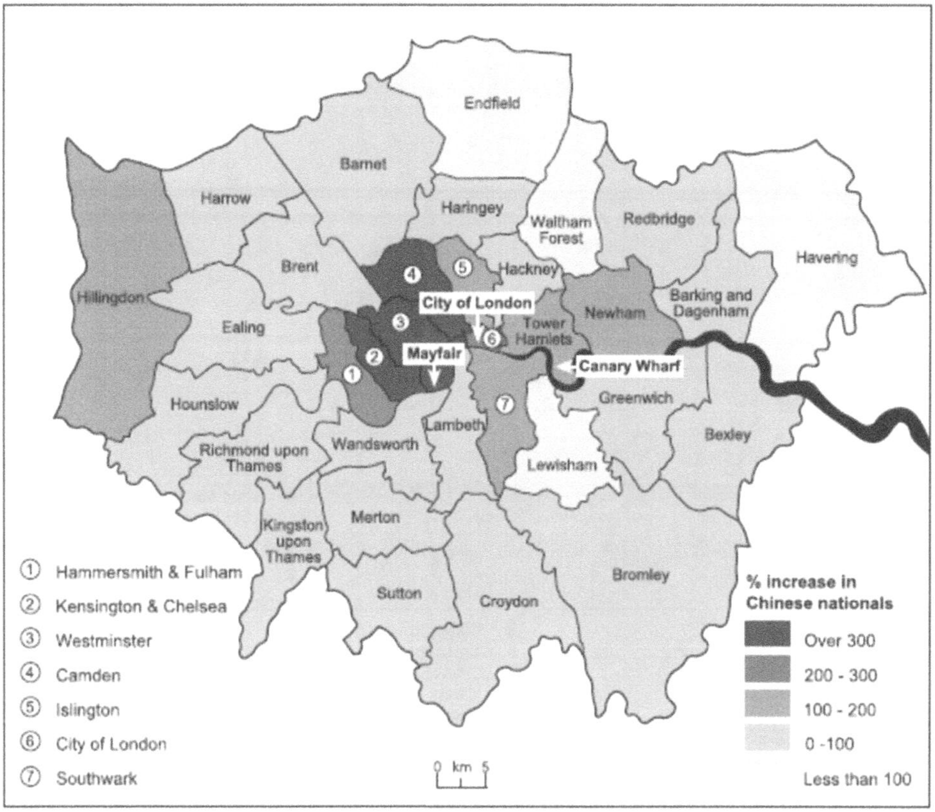

Figure 1. % increase in annual national insurance registrations for Chinese nationals by London Borough, 2002/3 – 2015/16.
Source: Department for work and pensions

financial services activity and therefore provide an indication that these migrants are working in financial and related professional services.

In addition to focusing on individual migrants, the impact of Chinese elite mobility to London working in financial services on London's urban form is also revealed by focusing on the location of Chinese financial firms operating in London. Previous research has clearly documented how waves of internationalisation have shaped both the working practices and associated urban form of London's financial district. This process is characterised by the expansion of international financial service firms from the historic centre of London's financial district close to the Bank of England (Pryke, 1991) eastwards to Canary Wharf (Daniels & Bobe, 1993), northwards and westwards in the case of hedge funds, asset managers and private wealth managers (Hall, 2017c).

Chinese financial services have not followed this trend of outwards geographical expansion beyond the historic centre of London's financial district. Indeed, Figure 2 shows that Chinese financial and related professional services are currently clustered closely around the Bank of England. The case of Chinese state owned commercial banks illustrates the preferences for these financial institutions to locate within the City of London most clearly.

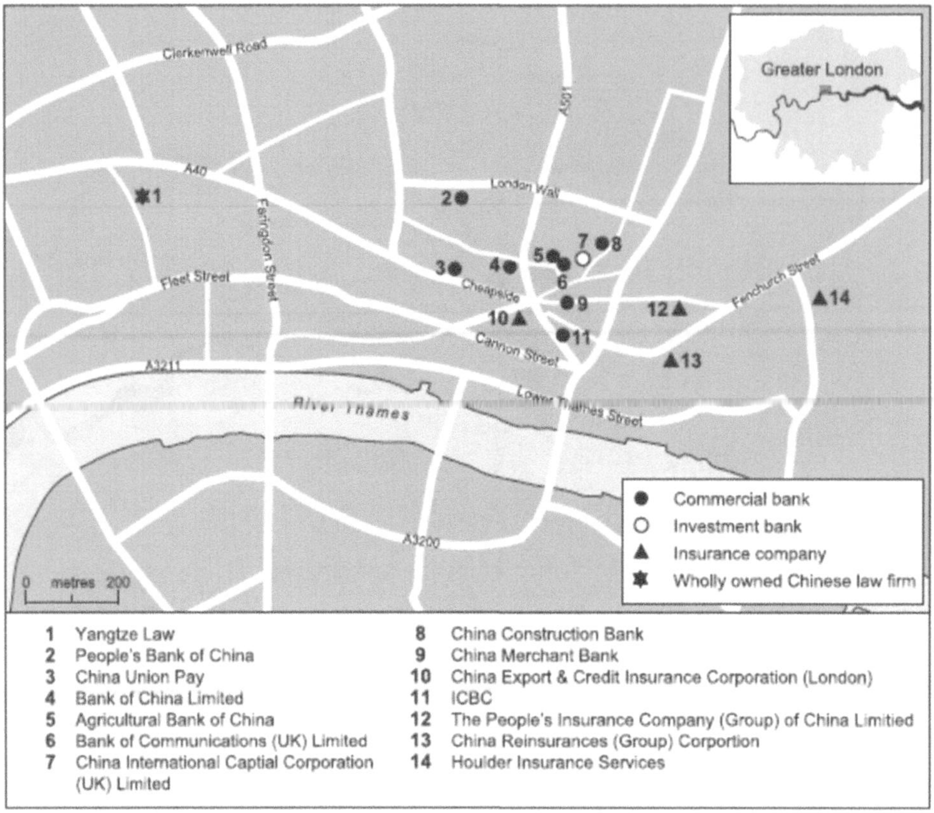

Figure 2. Location of main Chinese financial institutions operating in London.
Source: Author's research

In terms of bank internationalisation, the rapid and recent growth of London as a leading offshore RMB centre is reflected in the opening of branches by all four of China's biggest commercial banks in London from 2014 onwards (see Table 2). Prior to 2013, Chinese commercial state owned banks could not open branches in London because, in common with other non-EU financial institutions, Chinese banks could only establish subsidiaries rather than full branches in the UK. This reflected the cautious attitude of financial regulators in London to overseas banks in the fallout from the 2007–8 financial crisis (Ying, 2013). Bank subsidiaries were seen to pose less of a risk to systemic financial stability, because they had higher capital and liquidity requirements. The commitment to allowing Chinese banks in the UK to open branches was formalised in October 2013 at the Fifth UK-China Economic and Financial dialogue held in Beijing between the Chinese Vice Premier Ma Kai and the UK's Chancellor of the Exchequer George Osborne.[1] The first Chinese bank to be given such a licence was Industrial and Commercial Bank of China, which obtained its full UK banking licence in London at the end of 2014 and has been closely followed by the other big four state owned banks (see Table 2). The clustering of these banks around the Bank of England reflects their location in Beijing, where being close to the regulator, in this case the People's

Table 2. Big four Chinese commercial state owned banks operating in London, by tier 1 capital.

Bank	Date of founding in China	Headquarter location	Tier 1 capital $m	World Ranking by tier 1 capital (previous ranking)	Summary of international presence	Date granted UK bank branch license
Industrial and Commercial Bank of China	1984	Beijing	207,614	1 (1)	400 overseas branches in 39 countries	September 2014
China Construction Bank	1954	Beijing	173,992	2 (5)	76 overseas branches in Germany, Luxembourg, Vietnam, Sought Africa, Australia, US, South Korea, Singapore, Japan	January 2015
Bank of China	1912	Beijing	149,729	7(9)	100 branches in 36 countries	2015
Agricultural Bank of China	1951	Beijing	137,410	9(10)	Branches in Hong Kong, London, Tokyo, New York, Frankfurt, Sydney, Seoul and Singapore	2015

Sources: The Banker (2014) and bank annual reports

Bank of China (PBoC), is seen as being central to the learning of key market developments that will shape their organisational strategy and operations. As one banker observed:

> "You can draw a very tight circle around the Bank of England if you join the dots between all the Chinese offices in the City [of London]. And that is very Chinese – to be located as close as possible to the regulator"
>
> (Trader, Chinese bank, Chinese national, London, March 2015)

The clustering of Chinese professional service firms around the Bank of England is not limited to financial services. In the case of law, the growth of Chinese law firms has been dominated by mergers and joint operations between multinational law firms seeking to access the Chinese market and offer advice to Chinese clients seeking to invest outside of China, and their Chinese counterparts. For example, Baker & McKenzie, headquartered in Chicago, has developed a joint operation with the Chinese firm FenXun Partners. This allows Baker & McKenzie to access lawyers practising Chinese law which can only be undertaken by Chinese law firms. Responding to the growing demand from Chinese firms seeking to invest in the UK, the first wholly owned Chinese law firm was established in London in 2015 (Fortado, 2015). As Figure 2 shows, YangTse Law located close to London's financial district, reflecting the close relationship between financiers and lawyers when advising on investment into the UK.

Taken together, data on recent Chinese migration to London and the mobility of financial and related professional services firms into London provides an important first step in documenting the hitherto neglected growing significance of Chinese finance in London's financial district. However, in order to understand the rationale behind these urban geographies, in depth qualitative research is needed with Chinese migrants.

Understanding the motivations, experiences and trajectories of elite Chinese migration into London

Chinese finance in London is underpinned by two relatively distinct groups of elite Chinese financiers with different migration motivations, experiences and trajectories: first early career financiers building on their higher education credentials acquired in the UK; and second; financiers working on secondment from mainland China within Chinese commercial banks.

Educational routes into elite Chinese financial labour markets

The first group of elite Chinese financiers working in London's financial district are early career financiers who have stayed to work following completion of undergraduate or postgraduate degrees in the UK. As such, this groups builds on the more well studied mobility of Chinese higher education students to the UK (Brooks & Waters, 2011). Indeed, China sent the largest number of overseas non-EU students to UK higher education institutions in 2015–16 (the last year for which data is currently available, HESA, 2017). In common with the historic relationship between educational background and entry into elite financial labour markets (Jones, 1998), Chinese financiers strategise carefully about the institution they attended and the degree programme they studied, aiming for Oxford, Cambridge or another research-intensive Russell Group university, with a preference for quantitative based degree subjects, often in economics, mathematics, engineering or business studies. This reflects the ways in which these universities were seen as being able to confer the social and cultural capital needed to secure employment in London's financial district, as the following example demonstrates:

> "My parents had made it clear that travelling to the UK was a good idea, particularly as it would allow me to work in finance in London. But I had to go to a top ranked University to get the maximum benefit. So I studied economics at Warwick and every summer, applied for internships in London. I followed this with a Masters degree and hoped to use that and my work experience as a way of getting a finance job"
>
> (Associate, European investment bank, London, February 2015)

However, at another level, the case of Chinese migrants' educational routes into financial services work in London departs from established understandings of the relationship between educational background and financial services work in London that have, historically at least, often pointed to a relatively seamless transition from educational attainment at elite universities to employment in financial services (Augar, 2001). Rather, the experiences of young Chinese migrants reveal the challenges of converting highly valued educational credentials from UK universities into a career in financial services in London because of the continued lack of ethnic diversity within elite financial labour markets (see British Bankers Association, 2015; on the case of banking and Johanson & Vahlne, 2009; for a broader discussion of "otherness" within organisations). Indeed, progressing from university to a graduate level position in financial services, particularly in commercial financial services, was the exception rather than the norm amongst the individuals I interviewed. Different strategies were adopted to overcome this. Some graduates took a first position in retail banking, often in a back-office function in a provincial financial centre in which they had studied such as Birmingham or Edinburgh. For example, one banker reflected that

> "After my degree [first class economics degree from Birmingham], I applied to all the usual financial houses for their graduate schemes but received no offers. I think it was partly due to concerns about my English for writing reports. That was upsetting as my dream [of using UK education as a point of entry into financial services in the UK] had gone. So, I ended up taking a back-office job in Bank of China in Birmingham. It wasn't really what I wanted to do and wasn't very challenging but after two years I moved to their London operations and then moved from retail into their commercial operations."
>
> (Associate, European asset management, London, March 2015)

Even when Chinese migrants were employed in international banking brands, they were often located in parts of the labour market in which skills stereotypically associated with Chinese migrants were valued. Stereotypes based on ethnicity that assumed that Chinese migrants were *quantitative* financiers led to them being effectively limited to work in back office analytical roles, particularly if they sought to leave Chinese owned financial institutions and work for American or European firms, as the following banker reflected:

> "On paper, I have the perfect CV – a first [class degree] from Oxford and a masters in finance from London Business School. I started working for a Chinese bank but could only get hired here [a European bank] through a personal contact and even then, I'm stuck doing analytical work … my social networks and interests don't fit so well with the organization and I'm only being sponsored by another Chinese employee. I feel like they know I'm good at numbers but don't really look beyond that"
>
> (Analyst, European investment bank, London, March 2015)

Despite this separation, Chinese financial migrants were eager to use their experience as a stepping stone onto further mobility, often expressing a desire to gain international experience whilst they essentially waited for China's financial centres to develop in ways in which they could then bring their international experience to bear:

> I see London as part of my career but no more than that. I'm getting the experience I need here but will ultimately use that in another financial centre, possibly Singapore or maybe Beijing or Singapore when their financial systems are more mature. There I would be closer to home but still working in an exciting centre.
>
> (Analyst, UK bank, London, March 2015)

This discussion reveals how Chinese financiers working in London are typically at a much earlier career stage than many of the financial migrants and expatriates studied in existing work on financial elites (Beaverstock, 2004). As such, their career mobility is much more closely linked to their higher education choices, with their educational decision-making often explicitly linked to their financial career aspirations. However, their experiences of trying to secure employment within London's financial district reveals the challenge of trying to overcome the continued lack of ethnic diversity in elite financial labour markets to use the geographical career mobility of working in London as a way of advancing their careers by working in a leading IFC.

Secondments from China in London's financial labour market

The second group of Chinese migrants working in London's financial district are more senior bankers on secondment from mainland China. The financial institutions involved in this process are Chinese state owned commercial banks. Whilst "individual [Chinese] banks vary on their approach to staffing" (Vice president, Chinese state-owned bank, London February 2015), senior positions within these banks are typically filled through secondments from the Chinese head office with banks "not using the tier two visa for their own staff" (Head of China Desk, investment bank, London, April 2015). These individuals perform the role of "diplomat managers" seeking to secure personal career advancement through leading the implementation of regulations and expectations from Beijing, rather than focusing on market making in London. As one Chinese national banker in a Chinese state owned bank observed in June 2015 "I was head of the China desk and there were eight of us in a meeting and all of us were Chinese and none of us were recruited externally". A secondment typically lasts five years and this sets the time horizon within which these managers make decisions. As a result, a risk averse approach to business development predominates over stimulating market demand and establishing Chinese banks as legitimate service providers within London. This is because the incentive structure for these individuals prioritises work on the strategic internationalisation of Chinese banks, rather than developing links between these institutions and their international counterparts in London. Comments such as the following were typical amongst the Chinese bankers I interviewed:

> "Managers are here for around five years and that means they essentially act as diplomats rather than businessmen [sic]. They are more focussed on not messing up and making mistakes rather than the longer-term strategy of bank internationalisation."
>
> (Trader Chinese bank, Chinese bank, London, February 2015)

Recent work on the internationalisation of transnational service firms has increasingly demonstrated that corporate internationalisation relies not only on responding to client demand in new geographical markets, or following existing clients about new markets but instead focuses on stimulating demand by educating potential new customers into the value of using their services through a range of activities (see Suddaby & Viale, 2011). The reliance on expatriation within these banks in London, and the limited ability to make markets through stimulating client demand associated with this, limits the ability of Chinese banks to grow their client and product base beyond its extant focus on supporting Chinese small and medium sized enterprises operating in London (City of London, 2015). As one financial journalist summarised in London in February 2015:

> "Chinese banks should be able to create more products to get more clients. They need to face up to becoming a local bank, it is survival. It is a no brainer even to think about your [a Chinese bank] localisation strategy. So, the end of internationalization is localisation, how to become a local bank."

The limited ability of Chinese banks to make markets in this way is significant not only in terms of the career biographies of bankers working within them, but also in terms of the wider position of London within RMB internationalisation, revealing the on-going challenges of sustaining and developing London as an offshore RMB centre, as the following example demonstrates:

> "I think that has put them [Chinese banks] in a vulnerable position when they are here in an international market. … I'm feeling there is a gap between what the British Government wants them to do and what the Chinese banks can do in the market."
>
> (Capital markets lawyer, UK national, London March 2015)

This form of expatriation therefore differs in important ways from more well studied forms of expatriation from Europe and North America. The rationale is driven not by a desire to either create new market demand or to spread corporate best practice into a new geographical market (on which see Faulconbridge et al., 2009) but rather is closely entwined with political aspirations in Beijing aimed at developing London as a key node in the emerging RMB internationalisation network of IFCs (see also Hall, 2017a). Taken together, these different forms of elite migration show how elite mobility combines economic and political imperatives such that individual mobility patterns and practices are shot through with macro scale political and economic concerns. This finding is important because it offers a sympathetic critique of the existing literature on elite migration that typically emphasises the economic and personal motivations and experiences mobility, with less attention paid to the political dimensions.

Conclusions

In this paper, I have used the case of elite Chinese financial migration and expatriation to London's financial district to argue that comparatively neglected forms of elite financial mobility from beyond the Global North provide important insights into the changing geographical form and labour market practices within leading international financial centres. The main empirical contribution of the paper is to provide one of the first analyses to specify the nature of elite Chinese financial migration to London and locate this within calls to better understand the marked recent changes in Chinese migration to London more generally (Knowles, 2017). This is important because research has typically focused on lower skilled Chinese migration to London. As a result, the implications of the significant numbers of Chinese students studying at UK universities who are seeking graduate level employment in sectors such as finance and related professional services for elite labour markets and economic development within London (as revealed in this paper) and the rest of the UK remain poorly understood.

Whilst these findings have been developed through the unique case of elite Chinese migration into London, this research also contributes to wider agendas concerned with decentring understandings of the global economy (Robinson, 2005) by examining flows of elite mobility from rapidly developing economies beyond the Global North. In terms of elite labour market mobility, two implications are important. First, in common with earlier rounds of elite migration, Chinese migration is shaping London's urban form in important ways, not always following previous rounds of urban development stemming from internationalization in London's financial district. Here I support calls to more fully explore the intersection between mobility and cities, particularly in the case of Chinese mobility as this is playing an important but comparatively overlooked role in the ongoing economic development of London (Knowles, 2017). Second, elite Chinese financial migration to London challenges assumptions about the nature of migrants themselves, particularly in terms of their career stage and the rationales behind their mobility with personal career aspirations and political agendas from Beijing playing a more important role than organizational strategy which often dominates in existing understandings of elite migration and expatriation. Building on these findings is an important emerging research agenda for social scientists concerned with urban economies and their labour markets, particularly as migration becomes the subject of significant popular and political debate in many parts of the world economy, not least, the UK.

Note

1. Economic and Financial Dialogues have been held between monetary and finance ministers in UK and China since 2008 and typically include announcements concerning significant regulatory changes aimed at developing London's role within RMB internationalization.

Acknowledgments

An earlier version of this paper was presented at the Networks in Finance Workshop at Manchester Business School in December 2016. I would like to thank the audience at that event and Adam Leaver and Daniel Tischer for their invitation and engagement with my work. I would also like to thank Elaine Watts who prepared the maps, the individuals who gave up their time to be interviewed as part of the wider research project and the comments from four anonymous reviewers, David Bassens and Bas van Heur that have improved this version of the paper. Any remaining errors remain mine.

Disclosure statement

No potential conflict of interest was reported by the author.

Funding

This work was supported by the British Academy [MD130065].

References

Astbury Marsden. (2014, March 24). City of London lures Chinese speakers to handle surge of work. *Financial Times*. Retrieved July 22, 2016, from http://www.ft.com/cms/s/0/9951744a-b10b-11e3-bbd4-00144feab7de.html#axzz4EIBjUuN2

Atkinson, Rowland, Parker, Simon, & Burrows, Roger. (2017). Elite formation, power and space in contemporary London. *Theory, Culture and Society*, *34*(5–6), 179–200.

Augar, Philip. (2001). *The Death of gentlemanly capitalism*. London: Penguin.

Bassens, David, Derudder, Ben, & Witlox, Frank. (2010). Searching for the Mecca of finance: Islamic financial services and the world city network. *Area*, *42*(1), 35–46.

Bassens, David, Derudder, Ben, & Witlox, Frank. (2011). Setting Shari'a standards: On the role, power and spatialities of interlocking Shari'a boards in Islamic financial services. *Geoforum*, *42* (1), 94–103.

Beaverstock, Jonathan. (1996). Migration, knowledge and social interaction: Expatriate labour within investment banks. *Area*, *33*, 525–538.

Beaverstock, Jonathan. (2002). Transnational elites in global cities: British expatriates in Singapore's financial district. *Geoforum*, *33*(4), 525–538.

Beaverstock, Jonathan. (2004). Managing across borders': Knowledge management and expatriation in professional service legal firms'. *Journal of Economic Geography*, *4*(2), 157–179.

Beaverstock, Jonathan. (2005). Transnational elites in the city: British highly skilled migrants in New York City's financial district. *Journal of Ethnic and Migration Studies*, *31*(2), 245–268.

Beaverstock, Jonathan, & Hall, Sarah. (2012). Competing for talent: Global mobility, immigration and the City of London's labour market. *Cambridge Journal of Regions, Economy and Society*, *5* (2), 271–288.

Bohm, Anthony, Fallari, Marcelo, Hewett, Andrew, Jones, Sarah, Kemp, Neil, Meares, Denis, … Van Cauter, Kevin. (2004). *Vision 2020: Forecasting international student mobility a UK perspective*. London: British Council/IDP Australia.

British Bankers Association. (2015). *Diversity and inclusion in banking*. London: Author.

Brooks, Rachel, & Waters, Johanna. (2011). *Student mobilities, migration and the internationalization of higher education*. Palgrave MacMillan, UK.

City of London. (2015). *London RMB business volumes 2014 City of London, London*: Bourse Consult.

Daniels, Peter, & Bobe, J. M. (1993). Office building in the City of London: A decade of change. *Area*, *24*(3), 253–258.

Dominiczak, Peter. (2013, November 28). Most immigrants to the UK now come from China, figures show. *The Daily Telegraph*. Retrieved July 15, 2016, from http://www.telegraph.co.uk/news/uknews/immigration/10480785/Most-immigrants-to-the-UK-now-come-from-China-figures-show.html

England, Kim. (1994). Getting personal: Reflexivity, positionality and feminist research. *The Professional Geographer, 46*(1), 80–89.

Faulconbridge, James, Beaverstock, Jonathan, Hall, Sarah, & Hewitson, Andrew. (2009). The war for global talent. *Geoforum, 40*(5), 800–808.

Findlay, Allan, & Cranston, Sophie. (2015). What's in a research agenda? An evaluation of research developments in the arena of skilled international migration. *International Development Planning Review, 37*(1), 17–31.

Findlay, Allan, & Gould, W. T. S. (1989). Skilled international migration: A research agenda. *Area, 21*(1), 3–11.

Findlay, Allan, Li, F Lin, Jowett, A John, & Skeldon, Ronald. (1996). Skilled international migration and the global city: A study of expatriates in Hong Kong. *Transactions of the Institute of British Geographers, 21*(1), 49–61.

Florida, Richard. (2007). *The flight of the creative class.* New York: Harper Collins.

Fortado, Lindsay. (2015, May 31). Chinese-owned law firm seeks deals in west with London launch. *Financial Times.* Retrieved August 23, 2017, from https://www.ft.com/content/e05e603e-077f-11e5-a58f-00144feabdc0

Hall, Sarah. (2017a). Rethinking international financial centres through the politics of territory: Renminbi internationalisation in London's financial district. *Transactions of the Institute of British Geographers, 42*(4), 489–502.

Hall, Sarah. (2017b). Regulating the geographies of market making: Offshore renminbi markets in London's international financial district. *Economic Geography,* 1–20. doi:10.1080/00130095.2017.1304806

Hall, Sarah, & Wójick, Dariusz. (2018). 'Ground zero' of brexit: London as an international financial centre. doi: https://doi.org/10.1016/j.geoforum.2018.02.002

Hall, Ssarah. (2017c). *Global Finance: People, space and places.* London: Sage.

Higher Education Statistics Agency (HESA).(2017, August 24). Higher education student enrolments and qualifications obtained at higher education providers in the United Kingdom 2015/16. Retrieved from https://www.hesa.ac.uk/news/12-01-2017/sfr242-student-enrolments-and-qualifications

Iredale, Robyn. (2001). The migration of professionals: Theories and typologies. *International Migration, 39*(5), 7–26.

JLL. (2015). Jones Lang LaSalle London Luxury Quarter. Retrieved August 18, 2017, from http://www.jll.co.uk/united-kingdom/en-gb/research/283/london-luxury-quarter-global-destination-appeal

Johanson, Jan, & Vahlne, Jan-Erik. (2009). The Uppsala internationalization process model revisited: From liability of foreignness to liability of outsidership. *Journal of International Business Studies, 40*(9), 1411–1431.

Jones, Andrew. (1998). (Re)producing gender cultures: Theorizing gender in investment banking recruitment. *Geoforum, 29*(4), 451–474.

Kleibert, Jana, & Kippers, Lisa. (2016). Living the good life? The rise of urban mixed-use enclaves in Metro Manila. *Urban Geography, 37*(3), 373–395.

Knowles, Caroline. (2015). Who are the new Chinese migrants in the UK. *The Runnymede Trust.* Retrieved July 22, 2016, from http://www.runnymedetrust.org/blog/who-are-the-new-chinese-migrants-in-the-uk

Knowles, Caroline. (2017). Reframing sociologies of ethnicity and migration in encounters with Chinese London. *The British Journal of Sociology, 68*(3), 454–473.

Koser, Khalid, & Salt, John. (1997). The geography of highly skilled international migration. *Population, Space and Place, 3*(4), 285–303.

Krijnen, Marieke, Bassens, David, & Van Meeteren, Michiel. (2017). Manning circuits of value: Lebanese professionals and expatriate world-city formation in Beirut. *Environment and Planning A, 49*(12), 2878–2896.

Kunz, Sarah. (2016). Privileged mobilities: Locating the expatriate in migration scholarship. *Geography Compass, 10*(3), 89–101.

Lai, Karen. (2012). Differentiated markets: Shanghai, Beijing and Hong Kong in China's financial centre network. *Urban Studies, 49*(6), 1275–1296.

Ley, David. (2010). *Millionaire migrants.* Oxford: Blackwell-Wiley.

McDowell, Linda. (1992). Valid games? A response to Erica Schoenberger. *The Professional Geographer*, *44*(2), 212–215.
McDowell, Linda. (1998). Elites in the City of London: Some methodological considerations. *Environment and Planning A*, *30*(12), 2133–2146.
Moore, Fiona. (2016). City of sejourners versus city of settles: Transnationalism, location and identity among Taiwanese professionals in London and Toronto. *Global Networks*, *16*(3), 372–390.
O'Neill, Philip. (2008). Bringing the qualitative state back into economic geography. In T. J. Barnes, J. Peck, E. Sheppard, & A. Tickell (Eds.), *Reading economic geography*. London: Wiley.
Ong, Aihwa. (1999). *Flexible citizenship: The cultural logics of transnationality*. Durham, London: Duke University Press.
Pieke, Frank. (2007). Editorial introduction : Community and identity in the new Chinese migration order. *Population, Space and Place*, *94*, 81–94.
Pratt, Geraldine. (1993). Reflections on poststructuralism and feminist empirics, theory and practice. *Antipode*, *25*(1), 51–63.
Pryke, Michael. (1991). An international city going global: Spatial change in the City of London. *Environment and Planning D*, *9*, 197–222.
Robinson, Jennifer. (2002). Global and world cities: A view from off the map. *International Journal of Urban and Regional Research*, *26*(3), 531–554.
Robinson, Jennifer. (2005). Urban geography: World cities, or a world of cities. *Progress in Human Geography*, *6*, 757–765.
Ryan, Louise, & Mulholland, Jon. (2014). French connections: The networking strategies of French highly skilled migrants in London. *Global Networks*, *14*(2), 148–166.
Sassen, Saskia. (2000). *Cities in a world economy*. Thousand Oaks: Pine Forge Press.
Savage, Mike, & Williams, Karel. (2008). *Remembering elites*. Oxford: Wiley-Blackwell.
Saxenian, AnnaLee. (2006). *The new Argonauts: Regional advantage in a global economy*. Harvard: Harvard University Press.
Schoenberger, Erica. (1991). The corporate interview as a research method in economic geography. *The Professional Geographer*, *43*(2), 180–189.
Schoenberger, Eriva. (1992). Self-Criticism and self-awareness in research: A reply to Linda McDowell. *The Professional Geographer*, *44*(2), 215–218.
Sklair, Leslie. (2001). *The transnational capitalist class*. Oxford: Blackwell.
Smith, Michael Peter. (2001). *Transnational Urbanism: Locating globalization*. Oxford: Wiley-Blackwell.
Suddaby, Roy, & Viale, Thierry. (2011). Professionals and field-level change: Institutional work and the professional project. *Current Sociology*, *59*(4), 423–442.
The Banker. (2014, June 30). The banker top 1000 world banks 2014 rankings. Retrieved July 22, 2015, from http://www.thebanker.com/Top-1000-World-Banks/The-Banker-Top-1000-World-Banks-2014-rankings-WORLD-Press-release-For-immediate-release
The Financial Times. (2016, October 19). UK will protect free movement for bankers after Brexit. Retrieved February 19, 2018, from https://www.ft.com/content/eb27252d-94d8-328f-a0e3-76b890601484
Thrift, Nigel, & Leyshon, Andrew. (1992). In the wake of money: The City of London and the accumulation of value. In L Budd & S Whimster (Eds.), *Global finance and urban living: A study of metropolitan change*. London: Routledge.
Thrift, Nigel. (1994). On the social and cultural determinants of international financial centres: The case of the City of London. In S. Corbridge, R. Martin, & N. Thrift (Eds.), *Money, power and space* (pp. 327–354). Blackwell: Oxford.
Tomba, Luigi. (2004). Creating an urban middle class: Social engineering in Beijing. *The China Journal*, *51*, 1–26.
Töpfer, Laura-Marie, & Hall, Sarah. (2017). London's rise as an offshore RMB centre: State-finance relations and selective institutional adaptation. *Regional Studies*, 1–13. doi:10.1080/00343404.2016.1275538
Waters, Johanna. (2005). Transnational family strategies and education in the contemporary Chinese diaspora. *Global Networks*, *5*(4), 359–377.

Waters, Johanna. (2006). Geographies of cultural capital: Education, international migration and family strategies between Hong Kong and Canada. *Transactions, Institute of British Geography, 31*, 179–192.

Wildau, Gabriel, & Mitchell, Tom. (2016, December 11). Renminbi stalls on road to being a global currency. *Financial Times*. Retrieved August 24, 2017, from https://www.ft.com/content/e480fd92-bc6a-11e6-8b45-b8b81dd5d080

Xiang, Biao. (2003). Emigration from China: A sending country perspective. *International Migration, 41*(3), 21–48.

Ying, He. (2013). China's banks in London, centre for the study of financial innovation. Retrieved September 29, 2016, from http://csfi.org/files/chinabanklondonfinal.pdf

Z/Yen. (2016). *The global financial centres index 19 Z/ Yen,*Retrieved May 9, 2018, from www.longfinance.net/global-financial-centre-index-19/976-gfci-19-the-overall-rankings.html.

Follow the money: cultural patronage and urban elite geographies

David Bassens, Bas van Heur and Maëlys Waiengnier

ABSTRACT

Cultural patronage has since long been a domain where urban elites can showcase a sense of philanthropy. The paper develops a heuristic that starts from cultural patronage activities in order to trace urban elite households and their professional and residential characteristics. The heuristic is deployed in Brussels, Belgium, where we use cultural patronage listings from five key high-culture institutions as an empirical entry-point into urban elite geographies. Contrary to the common claim that current urban elites are global elites, our analysis shows a clear national reproduction of Brussels elites. Specific for Brussels is a remarkable level of nobility resilience, based on its more-or-less successful entanglements with *haute finance* as well as art market intermediary functions. Furthermore, the residential analysis suggests that elite reproduction also predominantly depends on socio-spatial segregation in inner-city neighborhoods, yet lacks morphological segregation strategies typical of enclave urbanism or super-gentrification generally found across global cities.

1. Introduction

Cultural patronage, i.e. patronage of the arts by wealthy individuals, has since long been a domain where economic elites can showcase a sense of philanthropy (Fasche, 2013; North & Ormrod, 1998; Ostrower, 1995). The wealthy support the construction of new cultural venues, sponsor cultural events, and contribute to financing the acquirement of new works of art. As the vast majority of the top cultural institutions are located in (larger) cities, cultural patronage is by and large a "city thing". More broadly, elites have historically played and continue to play a key role in supporting the arts and the wider cultural economy due to their high discretionary spending power. In this article, we build on but also depart from these various bodies of literature on urban elite culture. Instead of investigating the ways in which elites shape cultural organizations and urban cultural fields, we turn the coin over and use the material collected on cultural patronage to map the socio-spatial composition of elites. We are not interested in "culture" per se, but use empirical material on cultural patronage to gain a more sophisticated understanding of the occupational identities and residential locations of contemporary economic elites. In using cultural data to map economic elites, we thus contribute not just to the literature on culture and cities but also to political economic debates in urban studies. In particular, our paper aims to contribute to the emerging literature on the super-rich and related work in elite sociology (Daloz, 2010; Forrest, Koh, & Wissink, 2017; Hay & Beaverstock, 2016; Savage & Williams, 2008). First, we broaden the focus of elite research beyond the very wealthy to include others that hold significant economic power in a certain place. Second, our use of cultural institution patronage lists offers spatially refined data at the urban level, which is generally non-existent for the super-rich (but see Burrows, Webber, & Atkinson, 2017; DeVerteuil & Manley, 2017). Together, these advancements allow narrating an elite geography that is potentially more embedded and influential in shaping the urban political economy under study – in this case Brussels (Belgium).

Drawing on classic elite theory, a key problematic we seek to address in the Brussels case is whether there is a core and coherent group of elites that determines the direction of its political economy or whether elites themselves are internally differentiated with different groups of elites being powerful only in relation to particular policy arenas and societal institutions. In the context of research on cities, this was a key issue in the so-called community power debate in US urban political sociology of the 1950 and 1960s (Dahl, 1961; Hunter, 1953). Those studying growth machines (Logan & Molotch, 1987; Molotch, 1976; Rodgers, 2009) and urban regimes (Imbroscio, 2003; Mickey, 1997) posit that strong social relations between different types of elites generate the capacity to rule i.e. govern a city (Mossberger & Stoker, 2001). We sought to tackle this question by scrutinizing the occupational composition of social circles around key elite cultural institutions in Brussels. Subsequently, we mapped their residential locations as a heuristic to identify potential elite neighborhoods, since we were interested whether occupational variegation would result in distinct residential geographies. Moreover, we assumed that these residential geographies offered one explanation of elite social reproduction with neighborhoods operating as key sites for elite interaction (Atkinson, 2016; Butler & Lees, 2006). To achieve these goals, we constructed and relied on an original dataset using innovative methods that enabled us to gain a much more sophisticated and empirically detailed understanding of elites than is usually achieved in other studies. A well-known problem in elite research is the difficulty of gaining access to elites (Aguilar & Schneider, 2012). For this reason, we decided to rely purely on the collection of publicly accessible data to see if it would be possible to develop substantive conclusions on urban elites through a creative usage of easily accessible data: online (websites of the cultural institutions, White Pages address books, LinkedIn, and Google search) as well as offline (mapping and visual documentation of elite housing). This "ease of access" is a key strength and should make it possible to reproduce our approach in other settings.

The remainder of the article is structured as follows. Section two offers a brief review of the literature on urban elite geographies. We sketch the ways in which our empirical material both confirms and departs from existing literature on elites, focusing on recent research on the super-rich and research from the wider field of urban studies. The aim is to offer a strategic introduction to a conceptual vocabulary that can help us to interpret our empirical dataset with more analytical precision and nuance. Section three continues with a discussion of our methodology and the data collected. We explain our focus on a select number of "high culture" institutions as key entry points into elite composition in Brussels and offer a first presentation of our empirical material. Section four discusses the occupational variegation of elite networks around cultural patronage in Brussels. This is followed by section five, which provides an analysis of elite residential geographies. We conclude the paper by pointing to some of the strengths and limits of our methodological approach and identify productive directions for further research on urban elites.

2. Literature review: mapping urban elite geographies

2.1. *Methodological limitations of current research on elite geographies*

Within geography, the literature on the super-rich has for the last decade been the dominant strand of research for those interested in understanding elites (Beaverstock, Hubbard, & Rennie Short, 2004; Hay & Beaverstock, 2016). Such a focus on the super-rich has various methodological and conceptual implications for those wishing to draw on this body of work to study the urban geographies of elites.

First, even though the field is methodologically pluralizing, many contributions continue to (partially) draw on consultancy report definitions of high-net-worth individuals (HNWIs) (e.g. reports by Capgemini and RBC Wealth Management, Knight Frank, Wealth-X and UBS) to map out those with at least US$ 1 million in financial assets. The actual connections between economic elites and their urban environment therefore remain underdeveloped, which is to an important extent the result of various methodological difficulties that accompany the choice of working with the above definition. These reports and lists tend to be based on a combination of statistical analysis of household balance sheet data and surveys with more quali-quantitative biographical analysis of the assets of individuals and families. The measurement techniques underlying these reports are limited in various ways - not in the least the fact that the datasets are rarely publicly available and can thus not be verified - but being almost "the only game in town" most academic research refers to and builds on these reports and rankings. This limitation does not apply to the methodologically more advanced research on top incomes and "the one percent" by Anthony Atkinson, Emmanuel Saez, Thomas Piketty and others (Alvaredo, Atkinson, Piketty, & Saez, 2013; Wisman, 2013), which has gained quite some attention in the last years. This research relies on national tax statistics to analyze the long-run distribution of top incomes and explicitly investigates the share of national income and wealth taken by the top incomes. Both the reports and rankings and the top incomes research, however, suffer from a basic limitation in that the primary dataset and reference is the national level. Due to statistical limitations, no attention is paid to the geographical distribution of income and wealth within countries. Some of the consultancy reports do identify a list of cities in which billionaires or high-net-worth individuals are primarily concentrated, but the data underlying these rankings are particularly opaque and disconnected from the wider statistical analysis. The household survey data rely on the region as the smallest spatial unit for income, which in most cases is larger than specific urban areas, whereas data on wealth are generally only available at the country level. Keeping in mind these limitations, relevant work has been done by authors such as Sébastien

Breau, who has used Census microdata to map the geographical distribution of the top 1% across Canada. As he shows "the concentration of high incomes is predominantly an urban affair, with almost two-thirds of Canada's top 1% located in one of its five largest Census Metropolitan Areas (CMAs) (Toronto, Montreal, Vancouver, Ottawa-Hull and Calgary)" (2014, p. 20), which is approximately twenty percent higher than the share of the total population living in these five cities. Similarly, Danny Dorling and colleagues (2007) have investigated the spatial distribution of wealth in the UK using Family Expenditure Survey (FES) data, showing an increasing concentration of wealth over time in the South East of England. Using a different type of dataset, our paper contributes to this emerging body of critical research by drilling down to the most detailed level possible, namely to the level of individual elite households and their residential locations. Such use of non-aggregated data that also typifies investigative journalism invokes important questions about where the public responsibility of the researcher ends and the privacy of citizens begins. At the same time, the dominance of existing statistics very much works to veil the extreme levels of inequality that mark our society, and urgently requires creative circumventions to disclose possible anti-social elite behavior whereby elites mostly work to reproduce and isolate themselves at the expense of the larger population (cf. Dorling, 2014).

Second, when reviewing the geographical scope of the super-rich literature, operational definitions that center on HNWIs may well be justified when studying elite geographies in the most-connected global cities (e.g. London, New York, Singapore, Dubai). Yet, there is a danger that those interested in economic elite geographies beyond this global city shortlist draw on these operational definitions as objective, universal definitions to capture fundamentally different place-specific elite processes. Key global cities are prone to dynamics not representative for urban elite dynamics outside of these cities: they are marked by a larger concentration of extraterritorial command and control functions vested in financial and business services and their associated labor force (Bassens & van Meeteren, 2015) and a much tighter insertion of their property markets in global dynamics as a safe deposit box for accumulated wealth (Fernandez, Hofman, & Aalbers, 2016). The implication is that the super-rich literature offers an underdeveloped conceptual vocabulary that cannot fully grasp the role of economic elites in second or third-tier cities. In these cities, the position of global financial elites is much more ambiguous, while national, more rooted elites can be deemed at least as important as global financial elites (Andreotti, Le Galès, & Moreno-Fuentes, 2015).

These ambiguities very much apply to the case of Brussels, as the literature on its world-city formation has disclosed. Brussels' extraterritorial command and control functions (Elmhorn, 1998) are foremost diplomatic and political in nature, centered on the European Institutions and the NATO headquarters, regional representations, and a range of non-governmental institutions. These activities infuse the city with expats including Members of European Parliament, Commissioners, foreign journalists, and diplomats (Commissioner.Brussels, 2016). Lobbyists abound as well, especially around Place Luxembourg where the European Parliament and some of the Directorates-General are located, their presence being estimated at 25 000 (Corporate Europe Observatory, 2017). Indicative of more "classic" forms of world-city formation, Brussels also hosts a well-developed advanced producer service complex comprising the seats of foreign banks, financial infrastructure providers, management consultancy firms, advertising and public relations agencies, the Big 4 global accountancy firms, and international law firms. Taken together, the city's international functions are estimated to generate 17% of total employment in Brussels (i.e. 121 000 direct and indirect jobs; Commissioner.Brussels, 2016). The development of these functions has been a structuring force in the wider metropolitan area, reproducing and deepening existing forms of socio-spatial polarization (Kesteloot, 2000). But, as Swyngedouw and Baeten (2001) observed in the early 2000s, Brussels is also marked by a striking parochialism and limited levels of glocalization. Political agendas are generally dominated by linguistic and petty ideological strife, while global economic elites are usually excluded from the political arena. Despite the speculative practices of mostly local construction companies and financiers that seek to benefit from internationalization, there remains a general absence of glocal growth coalitions around the city's global functions. The general picture that emerges from the literature is hence one of a sociological disconnect between global elites operating in the international "Brussels Bubble" and a more rooted local elite that largely remains uncharted.

2.2 *Connecting the dots: geographical elite research and urban studies*

In our view, the interdisciplinary field of urban studies provides useful resources for more context-specific conceptualizations of urban elites and for designing methodologies to map their geographies beyond the global-city shortlist. First, the built environment is a viable entry-point as we are aware that specific spatial constellations and processes are associated with elite formation; paying attention to these local spatial orderings allows for nuanced relational definitions of eliteness. The current literature on the super-rich, however, tends to looks at "extreme" cases such as neighborhoods marked by super-gentrification, i.e. the gentrification of already gentrified areas by the super-rich (Butler & Lees, 2006) and elite enclave urbanism (Pow, 2011). Yet, even though socio-spatial segregation is a general (urban) phenomenon, it is also a process mediated by pre-existing social, political, morphological and other aspects of the urban regime (e.g. Burgers & Musterd, 2002). The consequence is that eliteness in a certain context is not necessarily associated with extreme luxury, conspicuous consumption, or strong morphological segregation, but involves more subtle socio-spatial strategies amidst relative residential mixity.

Second and as already indicated in the previous section, the literature on global and world cities warrants starting with specific extraterritorial economic functions as a heuristic to detect elite occupations and the associated social and work-related networks and spaces within the urban fabric. Quite often the financial sector and associated business services are taken as an entry-point to map out such geographies, including the leisure spaces of high-skilled professionals (Beaverstock, 2002, 2005) and their educational affiliations (Hall, 2008). These have proven to be powerful approaches, yet even in the case of global cities and even more so in second- and third-tier cities, the risk is that this may generate a partial reading of elites associated with specific (para)financial intermediaries in the global economy. Approaching eliteness from outside the realm of the economy sensu stricto may allow for a more decentered and open type of analysis.

A third helpful strand of literature focuses on the geographies of intermediaries. A prominent example is the recent attention granted to private wealth management services (Beaverstock & Hall, 2016; Beaverstock, Hall, & Wainwright, 2013). To date, however, these accounts have generated limited insight into the intra-urban geographies of those services, even though these geographies could be powerful indicators of elite neighborhoods as they tend to cluster close to wealthy clients. Beyond economic elite reproduction in the strictest sense through wealth management, it is therefore relevant to explore other ways in which intermediaries play a crucial role in the social reproduction of elites. In non-geographic terms, the literature on elite sociology sensitizes us to the social composition of urban elites, paying particular attention to the "internal workings of elite cultures" (Salverda & Abbink, 2013, p.3) as well as the differentiation of elites into groups with different and potentially competing lifestyles and interests. Here, both housing and cultural patronage can be considered as important expressions of elite distinction. The analysis of elite housing and the service infrastructure that supports elite real estate practices (e.g. real estate agents, private banking) goes much beyond the rather clichéd idea of elites living in super-gentrified urban enclaves. In fact, the ways in which elites live and use their housing as status symbols is very much shaped by cultural norms and thus geographically varied (Daloz, 2010). Similarly, research on cultural patronage emphasizes the differences between patronage cultures, especially between the US and Europe (Adloff, 2014). But with the exception of research on foundations, rarely any attention has been paid to the "internal workings" of elite patronage and how different kinds of intermediary actors shape the geographies of cultural patronage and elite involvement. Recent research on the cultural and creative industries has started developing such a relational approach to intermediaries (Jakob & van Heur, 2015), but to date no work has been done on the geographies of elite intermediaries in the field of culture.

2.3. *Tracing urban elites via cultural patronage*

If we thus want to capture the spatial distribution of economic elites within and between even smaller spatial units – ideally up to the urban neighborhood or even street level – we need to develop more experimental methods of data collection (cf. Froschauer & Wong, 2012). It is here that focusing on cultural institutions and cultural patronage activities may prove to be a viable entry-point. Up till now research on cultural patronage has investigated in detail the role of urban elites in the development of new cultural facilities as part of a wider urban development strategy (e.g. Strom, 1999) or – more relevant for our analysis – the role of cultural organizations as "elite ecologies" (Marcus, 1990). Elites are patrons of cultural organizations, because it is through this patronage that they develop and strengthen ties with other elites. In doing so, cultural organizations come to play an important role in elite status processes (Ostrower, 2002), and unpacking this dynamic may prove a fruitful methodology. A forthcoming paper by Accominotti, Khan, and Storer (in press), for instance, uses subscriptions at the New York Philharmonic to trace changes in elite composition. Although elite composition is related to occupation and residence, their empirical focus, however, is primarily historical (i.e. covering the period 1880–1910) and their research interests mainly sociological. Their work shows how socio-spatial segregation in the city is reflected in the seating patterns within the building and relates the broadening of the elite composition to the changing *raison d'être* of the cultural institution itself. Hence, whereas most of the established literature is "inward" looking (elite sociology and ethnography focusing mainly on elite behavior and motivations, and elite patronage studies focusing on the impact of elites on the organizational culture of arts organizations), we argue that this literature can also be drawn upon to look "outward". By identifying the elite composition of cultural patronage in a select number of cultural organizations, one can also map the professional occupations and residential geography of urban elites today and relate them to socio-spatial structures of inequality in the city. The next section details how this methodology was operationalized for the case of Brussels.

3. Methodology: tracing elites through cultural patronage

3.1. *Existing data on elite geographies in Brussels*

Even though much of the above-mentioned data issues also apply to Belgium, a handful of historians have used alternative data sources to map the residential geographies of elites in Brussels. The residential geography of bankers (from 1800–1970) has been compiled on the basis of information derived from commercial almanacs, high society and mundane telephone directories or financial/stock exchange collections listings (Bauwelinckx, 2011). Belgian nobility residences have been traced since the 18th century with help from historical travel guides containing hostels from the titled nobility (1768) and the *Carnet mondain* (the mundane telephone directory of the nobility, edited since 1983) (Debroux et al., 2007). Similar to the patronage lists used in this study, address lists from cultural associations – such as the *Cercle Artistique et Littéraire* (1847) or the *Salon National Triennal* – have been explored to trace the emergence of artistic elites at the end of the 19th century (Debroux, 2011). For the research underlying this paper, we built on this previous research and also took inspiration from more ethnographic "follow the thing" approaches as outlined in Marcus (1995). Even though our approach did not involve deep ethnographies, we decided to study and trace elites in the urban space of Brussels via the "thing" or "medium" of money, more specifically the money donated to cultural institutions by elite patrons.

3.2. *Building a sample of cultural patronage activities*

For the Brussels case, a strategic sample of five important cultural institutions and their patronage networks proved to be a viable empirical entry-point to start tracing the geographies of urban elites. All information on the cultural patrons is published on the websites and/or in freely available publications of the cultural institutions. This is an interesting observation as such, since elites are often understood to be very inaccessible due to their interest in hiding their wealth from public scrutiny. At the same time, the literature also shows that many elites have a particular interest in showing their wealth and in publicly "flaunting elitist connections" (Daloz, 2010, p. 96) – historically in the form of salon or club membership, but also art patronage.

It seems likely that the association of economic elites with high-status cultural institutions is an example of the latter. For us as researchers, this creates a very productive empirical entry-point into the world of urban elites and allows us to step away from bread-and-butter socio-economic data and to start acquiring and construing data sets in a more grounded way. The five cultural institutions are the following: Bozar, La Monnaie, Flagey, Queen Elisabeth Music Chapel, and Wiels. Bozar is the Center for Fine Arts, which organizes various cultural events (exhibitions, performances, and cinema). It is built in Art Deco style by Victor Horta (1861–1947), as part of Horta's urban development project on the "Mount of the Arts", close to the Royal Palace. La Monnaie is the Royal Theatre which also hosts the National Opera House. Within the building, the "Royal Box" and "Royal Salon" were frequented by the royal family and their guests. Nowadays sponsors and VIPs use these rooms as reception rooms. Flagey is a cultural institution covering multiple artistic disciplines, particularly music and cinema. It is located in the previous headquarters of the national radio and television buildings that, since the original owners left in 1974, has been giving residence to various cultural institutions. With Bozar, it is one of the prestige locations of the Queen Elisabeth Competition, an internationally renowned music competition. Founded by Queen Elisabeth of Belgium in 1939, the Queen Elisabeth Music Chapel is a public utility foundation that is nowadays Under the Honorary Chairmanship of Her Majesty Queen Paola. It is a higher-level training institution in music that co-produces events with a network of cultural partners in Belgium (Bozar, Flagey, La Monnaie). Wiels is a more recent contemporary arts center that opened its doors in 2008 with the idea of becoming "one of the leading institutions for contemporary art in Europe", housing national and international artists. It is located at the border of low-income neighbourhoods in the southwest of Brussels, in a renovated former brewery - one of the few examples of modernist industrial architecture in Brussels.

Three elements can already be outlined to motivate our "elite" appreciation of these institutions. First, all of the selected institutions are associated with "high culture" in the Brussels cultural milieu, which is usually understood as a key site for elite interaction. For the purpose of our argument it does not matter if we see these spaces of cultural patronage as inhabited by elites with shared aesthetic dispositions and intimate knowledge of high culture (Bourdieu, 1984) or if we reject this Bourdieuan argument and instead see these spaces more instrumentally as sites for elite networking by elites with or without a shared sense of the value of the arts (Ostrower, 1998). Second, three of the selected institutions, namely Bozar, La Monnaie, and Queen Elisabeth Music Chapel, show connections to Belgian royalty. Historically, these connections were more than symbolic in nature, but can be understood to reflect the dense Brussels-based social networks around Belgian royalty that supported investment in the Belgian (colonial) economy at the time. Entrance to these royal circles was paramount to get involved in Belgium's main holding company – the *Société Générale* – which monopolized industrial holdings in Belgium and the Belgian Congo (Oosterlynck, 2011, p. 489). It can be assumed that such legacies will play a role in marking the elitist nature of these institutions up to the present day. Third, and perhaps most evidently, patronage at these institutions comes at a cost that is generally not affordable to the middle classes. Yearly donations range between a minimum of €1000 and €12,500, and unlock perks such as discounts or free access to exhibitions, private events, invitations for openings, priority booking on concerts – in essence access to a more exclusive social circle centered on the particular cultural institution.[1]

3.3. *From patronage lists to elite occupations and residential spaces*

On the basis of the publicly available patronage lists of the five cultural institutions, three steps of data collection were performed. First, we integrated all lists and compiled one sample of a total of 552 elite households in Brussels.[2] This sample is therefore not exhaustive or representative in scope, but targets a certain kind of economic elites that have the cultural and economic capital to give financial support to cultural institutions. As we will show below, this sample bias allows us to identify and analyze an elite fraction that is much more rooted in place than the "footloose" global elites usually discussed in the literature on the super-rich. The size and pluralism of the sample also allows for a meaningful analysis of the composition of economic elites in Brussels and the relation between their residential patterns and existing socio-spatial divisions in Brussels. Second, for all households listed we looked up their residential addresses in the main online address directory of Belgium (White Pages), complemented by additional internet searches using Google. From a collected sample of 552 households, 38% of the addresses were located in the Brussels-Capital Region (i.e. 207 households). Another 15% were living outside of Brussels in Belgium and 3% outside Belgium (mainly France and Netherlands). For another 44%, no or insufficient data was available. This confirmed that we were essentially mapping an elite linked to Belgium with a concentration in the Brussels-Capital Region.[3] Third, patronage lists were combined with information about their professional occupation available through social media (LinkedIn) and additional Google searches. Information was retrieved for 85% of the sample (i.e. 468 households). The professions were codified into 13 categories of professional occupation according to the possibilities given by the entries, then gathered into types of sector.

4. Elite occupations in Brussels

By carefully analyzing the occupation composition of our cultural patronage sample, in this section we first tackle the theoretical question of whether one can identify a more or less stable core Brussels elite or, rather, whether elites in Brussels are internally differentiated with different networks of elites being powerful only in relation to specific societal domains. On the basis of our sample, we can identify which sectors and professions are represented in high cultural milieux. As our sample contains information about nationality, we then discuss to what extent these elites can be considered global or rather parochial elites marked by historical legacies. Finally, we point to the ways in which these elite networks of cultural patronage function as a market for elite services staffed by cultural and financial service intermediaries.

4.1 *The plurality of elites, but also a core...*

First and perhaps not surprising, the largest group in our sample (20%) consists of professions in the arts, culture and media (see Table 1). This category does not refer to

Table 1. Elite occupations in Brussels based on cultural patronage sampling.

Occupation	#	%
Arts, culture and media	110	20
Investment banking and financial holdings	47	9
Law	44	8
Nobility	42	8
Business services	39	7
Industry	41	7
Private banking, asset and wealth management	33	6
Construction and architecture	21	4
Real estate	24	4
Government (Belgium and European)	15	3
Education	8	1
Other	*44*	*8*
No data	*84*	*15*
Total	**552**	**100**

the artists themselves but principally to the art market and jobs related to marketing and communication. It gathers owners of art galleries, collectors, arts consultants, auction houses and some curators and publishers, all of whom have a professional interest in belonging to patronage clubs and many of whom seem to fulfil an important intermediary role: as experts advising other cultural patrons with more economic but less cultural, capital. Second, there is a strong presence of nobility within the sample (8%), mainly counts and barons. The nobility has a status of elite which has been perpetuated from the Ancient Regime until today and continue to occupy important functions in corporations or in the political sphere in Belgian society. Third, it could be assumed that occupations such as law, industry or *haute finance* (i.e. investment banking and financial holding activities) are traditionally associated with elites and these occupations indeed return in our sample, each of them counting for about 7 to 9%. Fourth, other sectors of activity not corresponding to the "traditional" elite suggest the presence of a "new" type of elites: the category business services (7%) includes professions in insurance, investment management and consultancy. Similarly, private banking, asset and wealth management (6%) is visible in our sample. Taken into account that the business services category also stretches beyond finance, finance as a more overarching category makes up about a fifth of the elite sample, putting it on equal footing with the arts, culture, and media category. Fifth, there is limited presence of real estate development (4 %) and construction and architecture (4 %) professions. Sixth and as discussed in the introduction to this paper, literature on urban growth coalitions and regimes assumes a strong link between business and governments, but in our sample the public sector is hardly represented: only 3% for the government and 1% for higher education. More precisely, it is a specific fraction of the public sector: university professors and functionaries working at the EU level. This does not mean that there are no connections between business and government in Brussels, but merely that these connections seem to be established and reproduced in settings outside of high culture. Finally, the "other or unknown" category represents a collection of diverse sectors: food, health, philanthropic activities, retail or unknown data and counts for 8% of the sample.

Our empirical evidence points to substantial actor diversity, but a diversity that is dominated by economic elites. Even within the "arts, culture and media" category, we observe – if not always economic elites themselves – primarily cultural intermediaries catering to an economic elite market. On a first reading of the data, this offers indicative support for the core elite thesis that different societal domains in Brussels are integrated by one "economically overdetermined" core of elite actors. A more nuanced reading of the data, however, also shows that there is only very limited actor overlap *between* the investigated cultural institutions: the vast majority of households are only patron of one cultural institution. This supports a pluralist perspective on elites as it suggests that even within the stratum of economic elites, there are various smaller social circles oriented towards particular cultural institutions. Having said that, certain actors and families do seem to perform a kind of "bridging function" both between the patronage networks of the various cultural institutions and between the cultural patronage networks and other elite networks such as business clubs. For example, the family Boël – one of the richest families in Belgium that developed its wealth in industry and later financial holdings – through one or more of its members is patron of the Bozar Center for Fine Arts as well as the Wiels center for contemporary art and has a history of intermarriage with other wealthy families, some of which also return in our sample (i.e. Solvay). Similarly Monique Bréhier, patron of Wiels, was until recently deputy chairman of the auction house Sotheby's Belgium and is also member of the Cercle de Lorraine, one of the main business clubs in Brussels. More advanced social network analysis of course would be needed to investigate these interlocking networks in detail, but this is not possible on the basis of the current dataset.

4.2 *Brussels and its parochial elites*

Our observations on the basis of our sample suggest that in Brussels urban elites are hardly footloose, but rather "parochial" elites, born and raised in Belgium. One would assume that the particular composition of elites would reflect the wider positionality of Brussels in political and economic networks at the Belgian, European, and global scale. Our observation is that this is in fact only to a very limited extent the case (cf. Baeten, 2001). Apart from the occasional French émigré (less than 3%), economic elites in Brussels are usually Belgians, confirming elite geographies in other European cities (Andreotti et al., 2015).[4] Employees of European and international institutions and the lobby complex are largely absent from our sample, indicating a divide in elite networks between a durable core of "national" economic elites and shifting networks of European elites with little interaction or exchange between them.

Nevertheless, one of the main insights of the super-rich literature – that we can observe a strong rise of financial and related professional elites and that these elites are footloose and globally oriented – is to some extent confirmed by our analysis. Finance, law (mainly business law), business and services and private banking, asset and wealth management indeed represent 30% of the total sample with a large share of investment banking, financial holdings, (international) law firms, and business service activities. For instance, elites take on high-level executive functions in the large Belgian-based financial service providers (e.g. AXA, KBC, BNP Paribas), in some global financial institutions (e.g. Credit Suisse, Barclays, ING), major energy holdings many of which have a French ownership connection (e.g. GDF Suez), and in a wide range of smaller boutique investment firms. Still, in contrast to what we can observe in other global cities, these networks remain firmly dominated by Belgian nationals. This confirms the hypothesis of an increased importance (although no dominance) of finance and related professional elites to the contemporary Brussels economy, but it also questions the often-claimed globality of these elites.

One element of both continuity and change is that some of the Belgian nobility (8% of our sample) is taking up new roles in the field of finance and investment. Besides employing rentier strategies, quite a few noble families are involved in financial holding and (investment) banking activities through connections with large Belgium-based financial institutions (e.g. AXA Insurance, Rotchild Private Bank Belgium, BNP Paribas). Historically, many noble families have had linkages with holding institutions (cf. Oosterlynck, 2011), but arguably the nature of financial services has changed significantly over the past few decades, becoming globalized and, more recently, financialized, thereby altering banking models completely (Erturk & Solari, 2007). The ongoing involvement of aristocratic families in these financial activities can be considered a sign of nobility resilience albeit under conditions of increased financialization that requires them to take on new professional positions. However, failed cases indicate that not all noble families understood the changing circumstances. The family de Selliers de Moranville in our sample, for instance, was one of the seven families with holdings in the original *Société Generale* that remained a major shareholder in what became Fortis Bank. Taking part in every round of share emissions, they and other aristocratic families took a serious financial blow when that bank went into failure in 2008, wiping out most of the family fortune.

4.3 *Elite services – wealth management, nobility and art market*

The applied "follow the money" approach is also powerful in detecting how elite circles function as a market for elite servicing. A first set includes a whole variety of wealth management services, which arguably is a key concern for elite reproduction (Beaverstock et al., 2013). A dense ecology of private banks, venture capital and private equity firms, asset management boutiques, wealth planning companies, and the likes are represented in our sample, indicating that their managers mix and mingle with other elites as part of their business strategy. The case of one patron of Wiels – Jean Berghmans – illustrates this point nicely. Having a London and Luxembourg history at Merrill Lynch, he started the asset management firm TreeTop, where he acts as CEO and self-proclaimed "wealth manager of rich Belgians". Other types of services are dedicated to (high end) real estate development, management, and investment. Some of these activities are global in scope with foreign ownership, but our sample also includes Belgian real estate investment trusts: such as Krest Real Estate Investments, which is owned by the Kandiyoti family that came under scrutiny in the Panama Papers scandal. This indicates that the elite networks also function as channels to allow capital switching into the built environment as a strategy to manage and maintain wealth. It remains an open question whether these circles are also used to influence the shape and direction of urban planning policies and development projects. While growth coalitions are important urban governance structures in Brussels (Baeten, 2001), the limited representation of government officials in our cultural patronage sample suggests that such typical growth coalition dynamics are less at stake here.

The largest category in our sample is that of professions in arts, culture and media. This category does include economic elites in their role of arts collectors: Pieter and Olga Dreesmann, collectors of Dutch Old Master paintings and members of one of the richest families in the Netherlands; or Herman J. Daled, an avid conceptual art collector who sold a part of his collection to the New York Museum of Modern Art in 2011. But by and large, this category consists of individuals and their companies acting as cultural intermediaries offering advice and expertise to other cultural patrons in the networks. This includes senior representatives from major arts auction houses such as Sotheby's and Christie's and various arts consultants. It also includes owners or directors of key galleries in Brussels, such as Xavier Hufkens, Albert Baronian, Roberto Polo and Harold t'Kint de Roodenbeke. In comparison to the other occupational categories and the overall setting (in Brussels as elsewhere) of economic elites being largely male elites, more women show up in the arts, culture and media category. This is not surprising, since this "cultural" side of the service industries – ranging from arts consultancy to curation and gallery ownership to arts education – has traditionally been a domain in which women have played a much more central role than in, for example, banking or business law. Finally, there is some evidence of overlap between elite networks and elite intermediaries, with a pronounced role for nobility in linking elite worlds. For example: Galerie Flore, a gallery for decorative arts in Brussels, is run by Flore De Brantes. She is the niece of Count Guy de Brantes, her uncle is the former president of France, Valéry Giscard d'Estaing, and she herself is married to Count Amaury de la Moussaye. An extreme example, perhaps, but other examples of cultural intermediaries connecting and literally embodying different elite fractions do show up in our sample.

5. Elite residential geographies

Following the discussion of elite occupations within our cultural patronage networks, in this section we use the same sample to investigate the residential geographies of elites. We contextualize our analysis by offering a very brief description of the historical geographies of elite housing in Brussels in order to highlight the local continuities between earlier and present-day elites. We then present and discuss the overall spatial distribution of elite housing across the Brussels-Capital Region, followed by a more detailed discussion of elite residential geographies in three key neighborhoods. For this last part, we organized field visits to (photographically) document the architectural and morphological properties of the identified residential addresses and their immediate neighborhood surroundings in order to get a clearer sense of elite residential preferences and the relation to the urban context of Brussels.

5.1 *Historical geographies of Brussels elites*

The residential patterning of elites in Brussels can usefully be understood as part of a much longer history of the urbanization of Brussels. With the development of industrial capitalism in the early 19th century, Brussels saw a huge demographic growth within the medieval city center and the immediate surroundings of the city. Municipalities developed basic road and zoning plans, but had little means to realize them, therefore leaving it largely to private initiative. Private land owners proposed speculative plans for their lands in the immediate surroundings of the city center (Deligne, 2011), while the nobility was eager to leave the center and move to the higher areas east of the center for fear of worker revolts as well as epidemics (Debroux et al., 2007). In these areas, new prestigious landscape elements such as gardens, squares and promenades were planned. The most emblematic example of this urban construction fervor by and for urban elites is the Quartier Léopold, a smaller part of the area that is currently the European Quarter. With construction activities funded by the *Société Générale*, its founders and shareholders were influential economic and political personalities (Deligne, 2011).

Following the independence of Belgium with Brussels as the new capital city and especially during the second half of the 19th century, the growth of the urban population continued. Not only did Brussels become an important financial center, the city also became increasingly attractive to artists with various opportunities to make a living in the city due to the proximity to and growth of urban elites. The cultivated bourgeoisie with its ideals of modernity constituted a wealthy clientele for artists (Debroux, 2011). Under the reign of King Leopold II, some of the wealth he had accumulated through his colonization of the Congo river basin in Africa was invested in prestigious building projects in Brussels. On the basis of spatial plans such as the *Plan d'ensemble pour l'extension et l'embellissement de l'agglomération bruxelloise* [Overall plan for the extension and beautification of the Brussels agglomeration], designed by Victor Besme, boulevards encompassing the municipalities of Schaerbeek and Saint-Josse-ten-Noode (east of the city center) and Ixelles and Saint-Gilles (south of the city center) were realized (Zitouni, 2010). By implicitly assigning each neighborhood to different functions or population groups, the plan reinforced existing socio-spatial divides. Smaller parcels were developed as middle class neighborhoods, in order to maximize the return on investment. Larger parcels were constructed in neighborhoods dedicated to the great bourgeoisie and aristocracy. The twentieth century saw a further expansion of the Brussels built environment. Because of the development of transport infrastructure combined with a very weak planning framework, many elites as well as the upper middle class suburbanized, a process far more pronounced than in other European cities (Halleux, Brück, & Mairy, 2002). At the same time, the socio-spatial structure inherited from the mid-19th century, with most elites living southeast of the medieval center, has remained important to this day (Debroux et al., 2007). This is also supported by our cultural patronage sample: a first quick scan shows that most of the identified elite households live outside of the core city center (the municipality nowadays called the City of Brussels) and tend to live in the wealthier urban municipalities to the southeast of this core. These residential patterns coincide with current socio-spatial divides in the city (Deboosere, Eggerickx, Wayens, & van Hecke, 2009) and thus show a remarkable continuity over time.

5.2 *Spatial distribution of elite households*

Analyzing the dataset in more detail, we can identify the exact spatial distribution of elite households in our sample. Figure 1 shows distribution of elite households by professional occupation across the Brussels-Capital Region.

The map shows a concentration of elite households in the southeast of the Brussels-Capital Region. Historically, the higher grounds located to the south and east of the

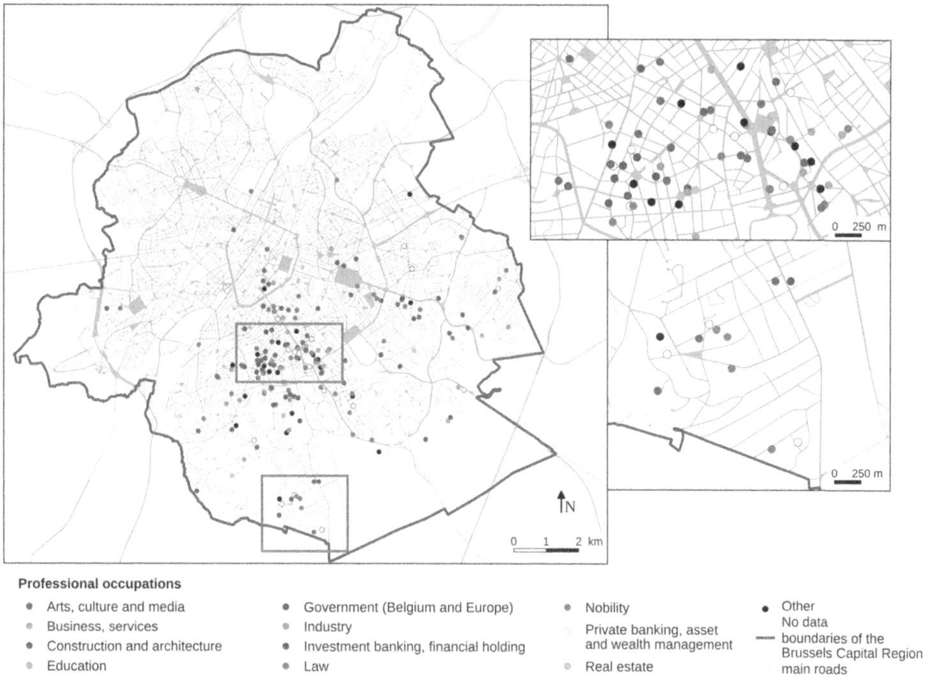

Figure 1. A profession-based overview of elite residential geographies in the Brussels-Capital Region.

medieval city have been the areas where the bourgeoisie resided, while the areas to the west of the city center were left to the poor, despite massive Hausmannization in the 19th century (Kesteloot, 2000; Vandermotten, 1997). These patterns have subsequently been reinforced after the Second World War due to internationalization, embodied in the arrival of guest workers and affluent workers of international organizations, but also have been superimposed by a concentric pattern due to suburbanization. The neighborhoods west of the city center are usually referred to as being part of the *croissant pauvre* [literally: the poor crescent]. This area hosts lower socio-economic status groups and is marked by a deteriorating, and overpopulated 19th-century housing stock. To the west of the "poor crescent" lies what is labelled in Brussels as the "working class belt", where the housing stock dates back to the 1930–1950s and neighborhoods are interspersed with a high density of retail activities (Romainville, 2015). Real estate prices are generally low and renovations few. We can note a very clear absence of elites into these two types of neighborhoods. Perhaps surprisingly, elites are also largely absent from currently gentrifying areas, such as the city center and central areas of the "poor crescent", which can be considered attractive locations. This suggests a permanence of elite residential locations in traditionally wealthier areas, which has been described as the continuous *embourgeoisement* of urban areas, involving repeated rounds of investment by economic elites in historically bourgeois neighborhoods (Andreotti et al., 2015). This is quite comparable to the oft-discussed case of London, with elites investing in "safe" areas that "have always maintained their value and prestige" and much less so in gentrifying areas in which rent gaps can be exploited (DeVerteuil & Manley, 2017, p. 1320).

In linking the occupational data on elites to their residential locations, we note that there is substantial spatial overlap between different occupational categories: elites in different sectors of the economy occupy the same set of neighborhoods in Brussels. It seems likely that these neighborhoods operate as "elite ecologies" with elites intermingling, frequenting the same shops, cafes and restaurants, and using the same wealth management and art market services as discussed in the previous section, but on the basis of our dataset and without further qualitative research this is impossible to say. Also, though not very pronounced, we can observe limited intra-elite differentiation when it comes to residential preferences. Three tendencies can be identified. First, there is a concentration of elite households working in traditional elite professions such as law or banking and "newer" professions such as business services along the prestigious axes of the *Avenue Louise* and the *Avenue de Tervueren*, respectively in the southeast and eastern part of the city. These large boulevards were developed in the second half of the 19th century for the urban elites of the time, linking the city with greener and less dense areas close to the forest (Deligne, 2011). Second, nobility to a large degree follows this same geography. This suggests a similarity in lifestyles and residential preferences between nobility and actors in law, banking and business services, which could be further investigated. A review of accounts on elite service clubs (Puype, 2004) and a preliminary desktop analysis of the geographies of private wealth management services in Brussels (following Beaverstock et al., 2013) suggest that critical supporting infrastructure is in place in these neighborhoods to cater to this clientele. Also, other research has shown that most international law firms tend to be concentrated along the *Avenue Louise* and that this has been the case since the beginning of the 20th century (Van Criekingen, Decroly, Lennert, Cornut, & Vandermotten, 2005). Third, households in the arts, culture and media category tend to live in the municipalities of Ixelles and Saint-Gilles, along the streets between *Avenue Louise* and *Chaussée de Charleroi*. This is a gentrified or gentrifying area undergoing a profound restructuring of the social and consumption landscape. The concentration of arts, culture and media elites in this area is in line with the wider literature in which artists are positioned at the forefront of gentrification processes: here, the elite households in the arts, culture and media category are not artists themselves, but can perhaps be understood as the most culturally inclined fraction of the elite prefiguring residential preferences of other elite fractions.

5.3 *Elite neighborhoods*

When we move down to the level of the neighborhood, new residential geographies become visible. Here we focus on three areas: the adjacent Brugmann-Lepoutre and Etangs d'Ixelles neighborhoods and the Vivier d'Oie further to the south https://monitoringdesquartiers.brussels/. We discuss the first two areas together, since they are neighboring areas and sociologically as well as morphologically very similar.

The Brugmann-Lepoutre and Etangs d'Ixelles neighborhoods are both part of the municipality of Ixelles to the south of the city center, which is one of the most expensive municipalities for housing in the Brussels-Capital Region. Although as elsewhere in Brussels, elites from various economic sectors are represented, one can observe a concentration of elites in the arts, culture and media category as well as the investment banking and financial holding category, especially in the Brugmann-Lepoutre neighborhood. Brugmann-Lepoutre borders the Saint-Gilles municipality that is much poorer, but also gentrifying and known for its cultural life with many cafes, venues and small-scale shops oriented towards niche consumption. Brugmann-Lepoutre and Etangs d'Ixelles are also both oriented towards urban leisure with many cafes and restaurants, but cater to a more upmarket audience than the neighborhoods in Saint-Gilles. Housing in these neighborhoods is by and large individual family housing from the late 1800s and the early 1900s in classicist, beaux-arts or eclectic architectural styles, in combination with large and luxurious apartment complexes built in the 1930s. Most of the individual houses have been transformed and split into multiple apartments, but (with the exception of elites living in the 1930s apartment complexes) almost all of the investigated elites own and occupy a full house; this is a first dimension of elite distinction in these neighborhoods.

By looking more closely at the housing of individual elite households, we can gain further insights into the residential preferences of elites and the ways in which they relate to their urban surroundings. A first example is the house on Avenue Molière 190 (see Figure 2), which appears in our sample as the residential location of Frédéric Samama, cultural patron of Bozar and Deputy Global Head at Amundi asset management in Paris, one of the largest French investment management companies. Stanford University educated and with a career in the US and France, Samama is one of the French émigrés that owns a house in Brussels. Located on the Avenue Molière that was constructed at the end of the 19th century, houses on this street were built around 1910 or in the interbellum. Some apartment complexes were built in the latter period, but most buildings are townhouses for individual families. Houses tend to have a small front garden with an iron fence; this is a second element of elite distinction in many of the urban neighborhoods in Brussels.

A second example is the apartment complex on Avenue Émile Duray 58 in the Etangs d'Ixelles neighborhood (see Figure 3). One of the patrons in our sample is linked to an apartment on this address, namely Claude de Selliers de Moranville. Patron of Flagey – which is within walking distance of the apartment – and Wiels, Claude de Selliers de Moranville, as discussed in section 4.2 of this paper, is member of one of the seven aristocratic families with holdings in the *Société Generale*, a major shareholder in what became the Fortis Bank that went into failure in 2008. Her apartment is part of a modernist apartment complex with art deco elements designed by the architect Adrien Blomme in 1927 (who also designed the brewery in which the Wiels is now located). A large entrance open to the public faces the street. Although the original design contains a concierge lodge, this did not seem to be used when we did our fieldwork (nor did we observe a concierge in other apartment complexes visited as part of the research). Inside, the complex contains a staircase and elevators for residents as well as service lifts that lead to the balconies on the rear of the building. Each floor contains two spacious apartments.

This basic architectural and morphological description of elite housing already shows a divergence from much of the literature on the super-rich and its discussion of elite housing with reference to notions such as super-gentrification or enclave urbanism. These conceptual tools are too blunt to really capture the dynamics of elite housing in most cities, and in any case in Brussels. Elites in our sample do live

Figure 2. The Samama residence on Avenue Molière 190 in the Brugmann-Lepoutre neighborhood.

in the wealthier areas in the city and a certain level of self-segregation takes place on the neighborhood level, but by and large there is limited "social concealment" of elites that involves strong forms of spatial closure (Atkinson, 2016): with the exception of simple iron gates no hard segregation barriers could be found. Also, due to the high levels of renters in Brussels, even the discussed neighborhoods are quite diverse: though clearly not accessible to lower incomes, the areas are also important residential and leisure spaces for students and workers with middle and upper middle class backgrounds and/or incomes. At most, it would be possible to identify spatial segregation between elites and "the rest" at the street level with some streets being unaffordable to anyone but economic elites, but with neighboring streets again more diverse in their social composition. The reasons for this may be multiple. Dessouroux (2012), for instance, suggests that the demand for security is subordinated by a desire by urban elites to maintain an urban lifestyle in close proximity to the city center, and with access to parks, public transport lines, and other common goods.

Figure 3. The apartment complex hosting the de Sellier de Moranville residence on Avenue Émile Duray 58 in the Etangs d'Ixelles neighborhood.

A third neighborhood marked by elite household concentration is Vivier d'Oie. Part of the municipality of Uccle, Vivier d'Oie is a residential and suburban-like neighborhood with one of the highest median income levels of the Brussels-Capital Region and in direct proximity to the *Forêt de Soignes*, a forest of more than 4000 hectares. Architecturally and morphologically, this is a completely different neighborhood. Consisting of freestanding villas or mansions, almost always with a large garden, this neighborhood has a population that is on average older than the population in the two other discussed areas and than in the Brussels-Capital Region. The urban lifestyle argument developed for those areas seems less applicable to this neighborhood, since here we do see stronger tendencies to hide wealth and to self-segregate through the use of hard segregation barriers. The most extreme example we found was the house on *Avenue des Eglantiers 18* that is owned by Pieter and Olga Dreesmann (see Figure 4), patrons of La Monnaie, collectors of Dutch Old Master paintings and other works of art, and members of one of the richest families in the Netherlands. Designed by the architectural office Marc Corbiau in 1991, this was the only residential location in our sample that would qualify as enclave urbanism, and even then, only on the level of the individual house. Hard segregation measures such as two layers of fences around the house and various cameras monitoring the plot have been installed here, most likely as an attempt to secure both the inhabitants of as well as various works of art in the house.

Most other examples of housing from the elites in our sample, however, are less extreme. Elements of self-segregation through spatial closure are common in this neighborhood, but are mostly limited to wooden fences and some physical distance between the street and the actual house. One example would be the house of Michel Delloye on *Avenue des Cytises 6* (see Figure 5), patron of Bozar and owner of the private equity fund Cytindus. By and large, however, the very socio-spatial form of the Vivier d'Oie neighborhood is shaped by a stronger demand for security and social concealment than is the case for the Brugmann-Lepoutre and Etangs d'Ixelles neighborhoods; the trade-off is that inhabitants in this area also remove themselves from an urban lifestyle with direct access to leisure, public transport and other common goods.

6. Conclusion

Witnessing the growing interest in, but also methodological limitations when tracing the whereabouts of elites in the literature on the super-rich, this paper has set out to develop a methodology around elite practices of cultural patronage to disclose a more fine-grained urban elite geography. By "following the money" behind cultural patronage listings, we were able to construct a more contextualized reading of eliteness in Brussels that has provided various insights into elite occupation and residential behavior. One of our central assumptions was that the retrieved elite would be reflective of the wider

Figure 4. The Dreesman residence on Avenue des Eglantiers 18 in the Vivier d'Oie neighborhood.

Figure 5. The Delloye residence on Avenue des Cytises 6 in the Vivier d'Oie neighborhood.

positionality of Brussels in European and global political economic circuits. The analysis has shown that this is only partially the case: while urban elites in Brussels take on positions in globally-oriented financial services and auxiliary services, salient connections to Belgian nobility circles are very much visible in our dataset. This suggests that social relations and hierarchies dating back to the Ancien Regime have remained surprisingly stable over time. Moreover, our elite sample is mostly composed of Belgian nationals, a condition that also marks the more rooted profile of upper middle-class elites in other European cities (Andreotti et al., 2015). A longitudinal analysis was not possible on the basis of this dataset, but the strong presence of nobility and their intertwinement with financial and art market intermediaries does suggest that – and this despite changing economic circumstances – the Brussels "parochial elite" consists of a somewhat closed group of actors.

This elite parochialism and associated rootedness, however, is not to be confused with an absence of globalization of their practices. In line with the results of Andreotti et al. (2015), elites in our sample show a rather nuanced and differentiated approach towards transnational mobility and "cosmopolitanness". Although our dataset shows the elites to be mostly born and raised in Belgium, many of these elites have occupationally become part of the financial and related service sectors operating on a global scale. Several of the listed elites also have educational experience abroad (mostly in the US) and/or have worked abroad for large multinational companies, again often in the finance and related sectors. This partially contradicts the observation of Swyngedouw and Baeten (2001) that Brussels shows a limited degree of glocal elite formation, since the local elites are clearly globalized in various ways. At the same time, Swyngedouw and Baeten's parochialism claim rings true more than ever, since we can observe a double disconnect: with the patronage elites seemingly disconnected both from the local political machinery (contra the theoretical assumptions of urban regime approaches) and from other elite circles such as the expat community in Brussels.

The question to what extent this group of elites can be understood as integrated or as plural (thus supporting either the core thesis or the pluralist thesis in elite theorizing) has to remain an open question. The data shows the overall dominance of economic elites and with certain key actors and families performing a "bridging function" between sectors. But the data also shows very limited actor overlap between the investigated cultural institutions, thus suggesting a more fragmented elite landscape in Brussels. The analysis of the residential geography of elites shows economic elites reproducing established socio-spatial divides in the city, yet diverging from extreme segregation typical of many other global cities. The reproduction of eliteness in Brussels seems to draw on more nuanced sociospatial strategies in which shared amenities in more-or-less central urban neighborhoods are a key characteristic. What we disclose here may be emblematic of a wider continental European variant of sociospatial distinction with an inner-city orientation and strong historical continuities; at the same time, even within Europe there is an enormous diversity to cities that should not be downplayed (Le Galès, 2002; Lenger, 2014). As typological or taxonomic exercises may work as theoretical straightjackets, the analysis of this diversity can perhaps be pursued more fruitfully through comparative urbanism tactics and their concomitant mindset of "thinking cities through elsewhere" (Robinson, 2016). More specifically, the insights about parochialism and subtle elite residential geographies in Brussels could indeed inform hypotheses in other contexts, even within places in the global city shortlist where the less conspicuous geographies of rooted elites remain under the radar of super-rich centered elite research.

A clear merit of the experimental methodology behind this study is that it is based on publicly available information, which should make it easier to apply this methodology to other urban settings and hence allow for nuanced comparative research. The approach pursued allows for highly detailed descriptions up to the neighborhood and even individual household level, something that would be very difficult to achieve with other methods and datasets (if only for reasons of privacy protection). Yet, while powerful, the approach also comes with a number of considerations. The main observation is that much of the interpretation of eliteness draws on the weight we give to socialization in high-culture circles. The involvement of financial and business elites next to those with noble roots indicates that, beyond being a site of old money reproduction, high culture institutions also serve as a meeting space for economic elites more broadly, in fact allowing us to start making claims about elite composition in relation to what we could expect to find in terms of (extraterritorial) political and economic functions of Brussels as global city. Yet, while we have strong contextual reasons to label the households in the sample as elites, we are fully aware that non-inclusion does not equal non-eliteness. In other words, while the sample offers us a clear view on elite composition, it may be the case that this sample is not representative of the wider elite community in Brussels. The implication is that, when seeking representativeness, the here documented insights would need to be corroborated through the building of datasets using other empirical entry-points (e.g. lobby networks, expat organizations, business communities, service clubs) or starting from altogether different spheres of social life and sports most notably (e.g. football, tennis, golf, etc.). Having said that, our approach shows the value of pursuing a more structured analysis of media(ted) data in urban studies and the study of elites. Media are key sites as these constitute platforms for economic elites to represent themselves and to be represented by others: in our study, this ranged from online address directories and the websites of the investigated cultural institutions to LinkedIn social media profiles, business websites, lifestyle magazines, local news media, and geneaology and family history websites. Follow-up research could build on this approach by compiling city- and domain-specific datasets on the basis of different kinds of media material and by conducting an in-depth discourse analysis of this material to identify key themes concerning the geographics of urban elites.

Even though the mapping of cultural patronage thus constitutes a viable elite heuristic, the method is best understood as an exploratory hypothesis-generating tool that requires follow-up research through a mixture of quantitative and qualitative methods. For instance, when expanding the institutional scope, cultural patronage and other elite activities could be modelled as a social network to detect social structures, which, when applied to longitudinal data could be studied in terms of their stability/dynamism. Further analyses could also broaden the data collection on household characteristics in terms of educational background. A Bourdieusian field analysis marshalling techniques such as multiple correspondence analysis (e.g. De Nooy, 2003) could subsequently shed more light on elite social structures undergirding these networks. Furthermore, the thus mapped social structures would invite a more thorough inquiry into the detected multilayered political-economic functions. Follow-up research could also study in depth how and to what extent these elite circles perform a role as a site for growth coalition formation. The strong professional overlap between nobility and financial factions invokes important questions about how key financial functions remain a bastion of the Brussels-based noble elite. Further, first evidence about the role of these elite circles as an art market in and of itself, but also as the place where a particular taste is socially construed would require a more in-depth study. Finally, while powerful in mapping elite neighborhoods, the study suggest that much can be learned by further developing ethnographic and morphological analyses in these places. Ethnography could be useful to find out whether residential behavior is a mere product of shared lifestyles, or whether residential choice is indeed key to understand elite socialization, for instance through educational and leisure facilities. As the Brussels case also shows residential continuity, important questions can be raised about the extent to which residential patterns can be explained by inheritance and intergenerational wealth conservation, a process guided by a range of private banking and wealth management facilities. If the power of a methodology is appraised not only in terms of the answers it provides, but also in the number of questions it invokes, then it appears that the cultural patronage approach presented in this article is a key instrument in the urban geographer's toolbox to start unpacking contemporary urban elite geographies.

Notes

1. Bozar: Amount unknown. La Monnaie / De Munt: 5000 Euro (Maecenas). Flagey: 1000 Euro (Great Friend) or 2800 Euro (Fellow). Queen Elisabeth Music Chapel: 1250 Euro (Patron) or 12,500 Euro (Sponsor). Wiels: 1000 Euro (Blue Member) or 2500 Euro (Gold Member). Median yearly income in the BrusselsCapital Region was ca. 15,800 euro for the

year 2014 (BISA, 2017a); in 2015 72% of the population earned less than 30,000 EURO, ca. 17% between 30,000–50,000 euro, and ca. 11% earned more than 50 000 (BISA, 2017b). This makes even the lowest fees in the database steep for the lowest quartiles and the highest fees unaffordable for the second quartile and may even put some stress at the lowest incomes of the upper decile (i.e. those earning more than 50,000 euro a year). An observation we cannot develop here is that compared to research on cultural patronage in the US (where the vast majority of research on patronage is conducted), these are very small amounts. Most likely, this has to do with radically different cultures of patronage between the two continents and the fact that economic elites in the US tend to be wealthier than their counterparts in Europe. A result is that in Brussels access to these elite settings is also affordable for upper middle classes. Our sample also shows this to be the case, although overall most households included in the sample are more usefully positioned on the elite end of the social spectrum. A possible explanation is that although the "entrance fee" of 1000–12.500 Euro is affordable in some ways, it gives access to a world in which further expenses seem almost obligatory (buying art, expensive travel, …).

2. Our empirical unit of analysis was the household, which we derived from the entries as provided in the patronage listings. Households can be individuals, but also – and often – couples. In the vast majority of cases, couples were listed with the first name of the male partner only (e.g. Monsieur et Madame Mickey Boël; Monsieur et Madame Jacques Zucker) and further desktop research showed that it was mostly the male partner to which the main income and wealth could be linked. This observation is in line with other research that shows wealth to be predominantly concentrated in male hands.
3. In principle, it would be possible that the majority of the 44% to which no address could be assigned live outside of Brussels or even Belgium. But there is no substantive reason to assume that this 44% shows a different spatial distribution than the other elite households.
4. The identification of individual actors in our sample as Belgian or non-Belgian is based on an "informed guess", since our dataset does not explicitly identify nationality. But on the basis of extensive LinkedIn and Google searches, we identified those actors as Belgian that followed their higher education in Belgium and/or that have spent most of their career in Belgium. This applies to the vast majority of all actors in our sample. French émigres, for example, have almost always studied in France and have a substantial career trajectory in France.

Acknowledgement

We would like to thank three anonymous referees for their valuable feedback on an earlier version of this paper. We also thank the Urban Geography editors and the special issue manager Susan Moore for their guidance throughout the revision process. The authors share equal responsibility for all remaining errors.

Disclosure statement

No potential conflict of interest was reported by the authors.

ORCID

David Bassens http://orcid.org/0000-0002-5368-7884
Maëlys Waiengnier http://orcid.org/0000-0003-4209-7245

References

Accominotti, Fabien, Khan, Shamus, & Storer, Adam. (in press). How cultural capital emerged in gilded age America: Purification and cross-class inclusion at the New York philharmonic. *American Journal of Sociology.*

Adloff, Frank. (2014). Philanthropisches Handeln in den USA und Deutschland: Zwischen Elitenreproduktion und Zivilgesellschaft [Philantropic Behavior in the USA and Germany: Between Elite Reproduction and Civil Society]. In Wolfgang Lauterbach, Michael Hartmann, & Miriam Ströing (Eds.), *Reichtum, Philanthropie und Zivilgesellschaft [Wealth, Philanthophy and Civil Society]* (pp. 181–197). Berlin: Springer.

Aguilar, Luis L.M., & Schneider, Christopher J. (Eds.). (2012). *Research amongst elites: Challenges and opportunities in studying up.* Farnham, UK: Ashgate.

Alvaredo, Facundo, Atkinson, Anthony B., Piketty, Thomas, & Saez, Emmanuel. (2013). The top 1 percent in international and historical perspective. *Journal of Economic Perspectives, 27*(3), 3–20.

Andreotti, Alberta, Le Galès, Patrick, & Moreno-Fuentes, Francisco J. (2015). *Globalised minds, roots in the city: Urban upper-middle classes in Europe.* Hoboken, NJ: Wiley-Blackwell.

Atkinson, Rowland D. (2016). Limited exposure: Social concealment, mobility and engagement with public space by the super-rich in London. *Environment and Planning A, 48*(7), 1302–1317.

Baeten, Guy. (2001). The Europeanization of Brussels and the urbanization of 'Europe'. Hybridizing the city: Empowerment and disempowerment in the EU district. *European Urban and Regional Studies, 8*(2), 117–130.

Bassens, David, & van Meeteren, Michiel. (2015). World cities under conditions of financialized globalization. *Progress in Human Geography, 39*(6), 752–775.

Bauwelinckx, Anne. (2011). Les espaces résidentiels du monde de la banque à Bruxelles (1822-1970). [The residential spaces of the Brussels banking world (1822-1970)]. *Les Cahiers de la Fonderie, 43,* 55–64.

Beaverstock, Jonathan V. (2002). Transnational elites in global cities: British expatriates in Singapore's financial district. *Geoforum, 33*(4), 525–538.

Beaverstock, Jonathan V. (2005). Transnational elites in the city: British highly-skilled inter-company transferees in New York city's financial district. *Journal of Ethnic and Migration Studies, 31*(2), 245–268.

Beaverstock, Jonathan V., & Hall, Sarah. (2016). Super-rich capitalism: Managing and preserving private wealth management in the offshore world. In Iain Hay & Jonathan V. Beaverstock (Eds.), *Handbook of wealth and the super-rich* (pp. 401–421). Cheltenham: Edward Elgar.

Beaverstock, Jonathan V., Hubbard, Phil, & Rennie Short, John. (2004). Getting away with it? Exposing the geographies of the super-rich. *Geoforum, 35*(4), 401–407.

Beaverstock, Jonathan. V., Hall, Sarah, & Wainwright, Thomas. (2013). Servicing the super-rich: New financial elites and the rise of the private wealth management retail ecology. *Regional Studies, 47*(6), 834–849.

BISA. (2017a) *Fiscale statistiek van de inkomens: Tabel 2.1.2.9: Mediaan total netto belastbaar inkomen der individuele aangiften 2005-2014.* [Median total net taxable income of individual returns]. Brussels Institutuut voor Statistiek en Analyse. Retrieved from http://bisa.brussels.

BISA. (2017b) *Fiscale statistiek van de inkomens: Tabel 2.1.1.3: Aantal aangiften per inkomensklasse: 2015.* [Number of individual returns by income class]. Brussels Instituut voor Statistiek en Analyse. Retrieved from http://bisa.brussels.

Bourdieu, Pierre. (1984). *Distinction: A social critique of the judgement of taste.* Cambridge, MA: Harvard University Press.

Breau, Sébastien. (2014). The occupy movement and the top 1% in Canada. *Antipode, 46*(1), 13–33.

Burgers, Jack, & Musterd, Sako. (2002). Understanding urban inequality: A model based on existing theory and empirical illustration. *International Journal of Urban and Regional Research, 26*(2), 403–413.

Burrows, Roger, Webber, Richard, & Atkinson, Rowland. (2017). Welcome to 'Pikettyville'? Mapping London's alpha territories. *The Sociological Review*, *65*(2), 184–201.

Butler, Tim, & Lees, Loretta. (2006). Super-gentrification in Barnsbury, London: Globalization and gentrifying global elites at the neighbourhood level. *Transactions of the Institute of British Geographers*, *31*(4), 467–487.

Commissioner.Brussels. (2016) Brussels-Europe: The Figures 2016. Downloaded via http://www.commissioner.brussels/i-am-an-expat/news/item/625-brussels-europe-in-figures

Corporate Europe Observatory. (2017) *Lobby Planet Brussels: The Corporate Europe Observatory Guide to the Murky World of Lobbying*. Retrieved from https://corporateeurope.org/

Dahl, Robert A. (1961). *Who governs? Democracy and power in an American City*. New Haven, CT: Yale University Press.

Daloz, Jean-Pascal. (2010). *The sociology of elite distinction: From theoretical to comparative perspectives*. London: Palgrave Macmillan.

De Nooy, Wouter. (2003). Fields and networks: Correspondence analysis and social network analysis in the framework of field theory. *Poetics*, *31*(5–6), 305–327.

Deboosere, Patrick, Eggerickx, Thierry, Wayens, Benjamin, & van Hecke, Etienne. (2009). The population of Brussels: A demographic overview. *Brussels Studies*. Synopsis nr. 3.

Debroux, Tatiana. (2011). Géographie de l'élite artistique bruxelloise à la fin du 19e siècle. [Geography of the Brussels artistic elite at the end of the 19th century]. *Les Cahiers de La Fonderie*, *43*, 31–37.

Debroux, Tatiana, Decroly, Jean-Michel, Deligne, Chloé, Galand, Michèle, Loir, Christophe, & Van Criekingen, Mathieu. (2007). Les espaces résidentiels de la noblesse à Bruxelles (XVIIIe–XXe siècle) [The residential spaces of the nobility in Brussels (18th-20th century)]. *Belgeo: Revue belge de géographie*, *4*, 441–452.

Deligne, Chloé. (2011). La ville et ses élites (1815-1915): Histoire d'une co-génération. *Les Cahiers de la Fonderie*, *43*, 10–17.

Dessouroux, Christian. (2012). Formes et géographie de l'enclavement résidentiel à Bruxelles. [Forms and geography of residential enclavement in Brussels]. *Les Cahiers Du Développement Urbain Durable*, *14*, 47–68.

DeVerteuil, Geoffrey, & Manley, David. (2017). Overseas investment into London: Imprint, impact and Pied-à-Terre urbanism. *Environment and Planning A*, *49*(6), 1308–1323.

Dorling, Danny. (2014). *Inequality and the 1%*. London: Verso.

Dorling, Danny, Rigby, Jan, Wheeler, Ben, Ballas, Dimitris, Thomas, Bethan, Fahmy, Eldin, ... Gordon, David, Lupton, Ruth. (2007). *Poverty, wealth and place in Britain, 1968 to 2005*. Bristol: Policy Press.

Elmhorn, Camilla. (1998). *Brussels: A reflexive world city*. Stockholm: Almqvist & Wiksell International.

Erturk, Ismail, & Solari, Stefano. (2007). Banks as continuous reinvention. *New Political Economy*, *12*(3), 369–388.

Fasche, Melanie. (2013). Making art history – Wealthy private collectors and contemporary visual art. In Iain Hay (Ed.), *Geographies of the super-rich* (pp. 171–185). Cheltenham: Edward Elgar.

Fernandez, Rodrigo, Hofman, Annelore, & Aalbers, Manuel. B. (2016). London and New York as a safe deposit box for the transnational wealth elite. *Environment and Planning A*, *48*(12), 2443–2461.

Forrest, Ray, Koh, Sin Yee, & Wissink, Bart (Eds.). (2017). *Cities and the super-rich: Real estate, elite practices and urban political economies*. London: Palgrave Macmillan.

Froschauer, Karl, & Wong, Lloyd. (2012). Studying immigrant entrepreneurs: From ordinary to big players in the transnational capitalist class. In Luis L.M. Aguilar & Christopher J. Schneider (Eds.), *Researching amongst elites: Challenges and opportunities in studying up* (pp. 159–178). London: Routledge.

Hall, Sarah. (2008). Geographies of business education: MBA programmes, reflexive business schools and the cultural circuit of capital. *Transactions of the Institute of British Geographers*, *33*(1), 27–41.

Halleux, Jean-Marie, Brück, Laurent, & Mairy, Nicolas. (2002). La périurbanisation résidentielle en Belgique à la lumière des contextes suisse et danois: Enracinement, dynamiques centrifuges et régulations collectives [The Residential Peri-Urbanization in Belgium in Light of the Swiss and Danish Contexts: Rootedness, Centrifugal Dynamics and Collective Regulations]. *Belgeo, 4* (4), 333–354.

Hay, Iain, & Beaverstock, Jonathan V. (eds.). (2016). *Handbook on wealth and the super-rich.* Cheltenham: Edward Elgar.

Hunter, Floyd. (1953). *Community power structure.* Chapel Hill, NC: University of North Carolina Press.

Imbroscio, David L. (2003). Overcoming the neglect of economics in urban regime theory. *Journal of Urban Affairs, 25*(3), 271–284.

Jakob, Doreen, & van Heur, Bas. (2015). Taking matters into third hands: Intermediaries and the organization of the creative economy. *Regional Studies, 49*(3), 357–361.

Kesteloot, Christian. (2000). Brussels: Post-fordist polarization in a fordist spatial canvas. In Peter Marcuse & Ronald van Kempen (Eds.), *Globalizing cities: A new spatial order?* (pp. 186–210). Hoboken, NY: Wiley.

Le Galès, Patrick. (2002). *European cities: Social conflicts and governance.* Oxford: Oxford University Press.

Lenger, Friedrich. (2014). *Metropolen der Moderne: eine Europäische Stadtgeschichte seit 1850 [Modern Metropolises: a European Urban History since 1850].* München: C.H. Beck.

Logan, John R., & Molotch, Harvey L. (1987). *Urban fortunes: The political economy of place.* Berkeley, CA: University of California Press.

Marcus, George E. (1990). The production of European high culture in Los Angeles: The J. Paul Getty trust as artificial curiosity. *Cultural Anthropology, 5*(3), 314–330.

Marcus, George E. (1995). Ethnography in/of the world system: The emergence of multi-sited ethnography. *Annual Review of Anthropology, 24*(1), 95–117.

Mickey, Lauria (Ed.). (1997). *Reconstructing urban regime theory: Regulating urban politics in a global economy.* Thousand Oaks, CA: Sage.

Molotch, Harvey L. (1976). The city as growth machine. *American Journal of Sociology, 82*(2), 309–332.

Mossberger, Karen, & Stoker, Gerry. (2001). The evolution of urban regime theory: The challenge of conceptualization. *Urban Affairs Review, 36*(6), 810–835.

North, Michael, & Ormrod, David (Eds.). (1998). *Art markets in Europe, 1400-1800.* Farnham, UK: Ashgate.

Oosterlynck, Stijn. (2011). From national capital to dismal political world city: The politics of scalar disarticulation in Brussels. In Ben Derudder, Michael Hoyler, Peter J. Taylor, & Frank Witlox (Eds.), *International handbook of globalization and world cities* (pp. 487–496). Cheltenham: Edward Elgar.

Ostrower, Francie. (1995). *Why the wealthy give: The culture of elite philanthropy.* Princeton, NY: Princeton University Press.

Ostrower, Francie. (1998). The arts as cultural capital among elites: Bourdieu's theory reconsidered. *Poetics, 26*(1), 43–53.

Ostrower, Francie. (2002). *Trustees of culture: Power, wealth, and status on elite arts boards.* Chicago: The University of Chicago Press.

Pow, Choon-Piew. (2011). Living it up: Super-rich enclave and transnational elite urbanism in Singapore. *Geoforum, 42*(3), 382–393.

Puype, Jan. (2004). *De Elite van België: Welkom in de Club [The Elite of Belgium: Welcome to the Club].* Antwerpen: Van Halewyck.

Robinson, Jennifer. (2016). Thinking cities through elsewhere: Comparative tactics for a more global urban studies. *International Journal of Urban and Regional Research, 40*(1), 3–29.

Rodgers, Scott. (2009). Urban growth machine. In Rob Kitchin & Nigel Thrift (Eds.), *International encyclopedia of human geography* (pp. 40–45). Amsterdam: Elsevier.

Romainville, Alice. (2015). La production capitaliste des logements à Bruxelles. *Carnets de géographes,* 8.

Salverda, Tijo, & Abbink, Jon. (2013). Introduction: An anthropological perspective on elite power and the cultural politics of elites. In Jon Abbink & Tijo Salverda (Eds.), *The anthropology of elites: Power, culture, and the complexities of distinction* (pp. 1–28). London: Palgrave Macmillan US.

Savage, Mike, & Williams, Karel (Eds.). (2008). *Remembering Elites*. Hoboken, NY: Wiley-Blackwell.

Strom, Elizabeth. (1999). Let's put on a show! Performing arts and urban revitalization in Newark, New Jersey. *Journal of Urban Affairs*, *21*(4), 423–435.

Swyngedouw, Erik, & Baeten, Guy. (2001). Scaling the city: The political economy of glocal development - Brussels' Conundrum. *European Planning Studies*, *9*(7), 827–849.

Van Criekingen, Mathieu, Decroly, Jean-Michel, Lennert, Moritz, Cornut, Pierre, & Vandermotten, Christian. (2005). Local geographies of global players: International law firms in Brussels. *Journal of Contemporary European Studies*, *13*(2), 173–187.

Vandermotten, Christian. (1997). L'impact du contexte politique et institutionnel sur les choix urbanistiques et économiques : la Région de Bruxelles-Capitale. [The impact of the political and institutional context on urbanistic and economic choices]. In Paul Claval & André-Louis Sanguin (Eds.), *Métropolisation et Politique* (pp. 107–123). Paris: L'Harmattan.

Wisman, Jon D. (2013). Wage stagnation, rising inequality and the financial crisis of 2008. *Cambridge Journal of Economics*, *37*(4), 921–945.

Zitouni, Benedikte. (2010). *Agglomérer: une anatomie de l'extension bruxelloise (1828-1915) [Agglomerate: an Anatomy of the Brussels' Extension]*. Brussels: VUB Press.

Websites

BOZAR. (n.d.). Retrieved March 15, 2016, from http://www.bozar.be/en

Chapelle Musicale Reine Elisabeth. (n.d.). Retrieved March 15, 2016, from http://musicchapel.org/

De Munt. (n.d.). Retrieved March 15, 2016, from http://www.demunt.be/nl/

Flagey. (n.d.). Retrieved March 15, 2016, from http://www.flagey.be/

LinkedIn. (n.d.). Retrieved from http://www.linkedin.com

Monitoring des Quartiers. (n.d.). Retrieved from http://monitoringdesquartiers.brussels

WIELS | Contemporary art centre. (n.d.). Retrieved March 15, 2016, from http://www.wiels.org/en/

Witte Gids. (n.d.). Retrieved from http://www.wittegids.be/

Index

Page numbers in **bold** refer to tables and those in *italic* refer to figures.

For Product Safety Concerns and Information please contact our EU representative GPSR@taylorandfrancis.com Taylor & Francis Verlag GmbH, Kaufingerstraße 24, 80331 München, Germany